GET RICH WITH
APPS!

Your Guide to Reaching More Customers and Making Money *NOW*

JESSE FEILER

Mc
Graw
Hill

New York   Chicago   San Francisco   Lisbon   London   Madrid   Mexico City
Milan   New Delhi   San Juan   Seoul   Singapore   Sydney   Toronto

While the author and publisher have done their best to ensure that the screen shots presented in this book are current at the time of printing, the reader must be aware that due to the ever-evolving technology of the medium it is impossible to gurarantee the accuracy of every screen shot once the book has been published.

Readers should know that online businesses have risk. Readers who participate in marketing and selling online do so at their own risk. The author and publisher of this book cannot guarantee financial success and therefore disclaim any liability, loss, or risk sustained, either directly or indirectly, as a result of using the information given in this book.

Contents

Preface

Rarely has a technology taken off as quickly as Facebook and iPhone applications, or *apps*, have. It's safe to say that you will be using apps either as a developer, a marketer, a business owner, or an end user. The only question is whether you will recognize their potential to change the way people use computers and the Internet—and their potential to help you make money. This book introduces you to the world of apps and how you can profit from that world. Even if you have used apps on Facebook or your iPhone/iPod touch, have you thought about how they work and where they come from? More important, have you thought about your role in this rapidly expanding world?

Get Rich with Apps! will show you how to look at apps in a new way. Don't just use them; watch how you and others use them. See how people interact with them, what works, and what doesn't. Then look for that sweet spot where you can make your own special (and profitable) contribution. In the first part of the book you'll peek behind the curtain of apps and learn how they work. You'll find details specific to both iPhone and Facebook. In the second part of the book you'll find basic strategies for making money from apps—you can pick and choose among them or even combine two or more to create your own unique app or set of apps.

Facebook and iPhone apps represent a new era in software. For a variety of reasons (described in Part 1), they are structurally different than old-fashioned application programs. The development times and costs for these new apps are only a small fraction of those associated with software products of even five years ago. In late 2009, an article in *Business Week* reported that it cost $2,000 to $10,000 to develop an iPhone app from scratch. As you will see in this book, there are simpler, faster, and less expensive ways to build iPhone apps if you use the iPhone Web app model. And Facebook apps are less expensive still.

This radical restructuring of the cost of software development has changed the way forward-thinking people approach the idea of software development. With such low costs, it's possible to develop apps (or have them developed) for even the smallest business. Developing an app can often be less expensive than placing an ad in a magazine. No longer do you have to be a risk taker or adventurer to develop iPhone and Facebook apps—not doing so is a bigger risk.

Apps even form the core of Apple's new iPad, thus providing a common development platform for iPhone, iPod touch, and iPad. Most iPhone apps run on iPad without any changes. Perhaps most important, the features that make iPhone apps attractive and useful to people are also present on iPad. Among the most important of those features to end users, developers, and entrepreneurs is still the ease and speed of development as well as the low cost to both developers and end users. It is likely that iPad will not be the last new device built on a framework of apps.

Fortunately, most of your competitors are probably still thinking about moseying up to the starting gate. This book helps you understand what's happening and how you can profit from it.

Although the details of implementing iPhone and Facebook apps change rapidly, their basic structures are remarkably stable. If you are a developer, you will want to keep up with the details, but the big picture described in this book is unlikely to alter much. Despite the massive changes and rapid growth in this area, the big picture would be recognizable to the pioneers from the olden days of apps (two or three years ago).

The rapid changes in this world are striking, but now is the time to understand what's happening and what role you can play. It is at the time of rapid growth of a new technology that some of the biggest opportunities are found: the field is wide open.

Just as a great land rush opened settlement of the western territories in the United States, the great app rush is opening a new way of developing and using software and the Internet. It's time to stake your claim.

In Part 2 of this book you'll find strategies for getting in on the great app rush. Once you understand apps and what you can do with them, you'll see how to sell them, promote your business with them, and use them to drive traffic to your brick-and-mortar and website locations. The large number of apps and their users provide a range of additional opportunities—including the design of app layouts, infrastructure for the back ends of apps, co-marketing opportunities, and the provision of services to app users.

■ ■ ■ Find More Information

It's important to keep up with changes in the world of apps. To do so, visit northcountryconsulting.com, where you can find updates to the information in this book and sources for additional material. You can also sign up for a newsletter to stay informed on the latest news and developments.

■ ■ ■ Acknowledgments

Thanks to the many people who have helped make this book possible. Carole Jelen at Waterside Productions helped create the idea for the book. At McGraw-Hill, Donya Dickerson, Tania Loghmani, Fiona Sarne, and Susan Moore provided great support and suggestions. Thanks also to the Clinton Essex Franklin Library System (CEFLS), Ewa Jankowska, and Betty Brooks for the CEFLS iPhone screen shots.

PART 1

Introducing the World of Apps

1

The Future of Apps 3.0
for You and Your Business

APPS 3.0 ARE SURGING like a gigantic wave. Are you waiting to be swamped, or are you preparing to ride the wave to personal profit and business growth?

The story of apps can be told in many ways. How many people do you know who will admit to not knowing about Facebook or iPhone, at least in general terms? "Create a Facebook Page" has been added to the to-do list for everyone from political advisors to realtors, large and small businesses, nonprofits, and individual users from teens to seniors. (Facebook has even developed a policy for handling the pages of people who have passed away.) On the iPhone side, "There's an app for that" is listed in *The Yale Book of Quotations* (Fred R. Shapiro, editor) as the third most notable quotation of 2009.

Alongside the anecdotal stories of Apps 3.0, another one is told in numbers. Here are some that you'll likely notice first:

- More than 50 million iPhones were sold in the product's first two years.

- There are more than 350 million active Facebook users, half of them logging on every day.
- There are more than 140,000 third-party apps for iPhone and more than three billion downloads from Apple's iTunes App Store.
- There are more than 350,000 third-party Facebook apps, and more than 70 percent of Facebook users use third-party apps each month. More than 250 third-party Facebook apps have more than one million monthly active users.

On the other side of the coin, not all statistics in the Apps 3.0 world involve large numbers:

- Facebook apps are free.
- iPhone apps are free or relatively inexpensive, with common price points of $4.99 and $9.99, although some cost $29.99 or more.
- The estimated cost to develop a sophisticated iPhone or Facebook app, such as Twitterific (iPhone) or those used by the Obama Campaign, can be anywhere from $10,000 to $150,000. Basic apps can cost a tenth of that—or even less.
- Developers commonly charge hourly rates of $50 to $150.

Finally, you may see these numbers:

- The Facebook app for iPhone is consistently one of the most downloaded free apps.
- People who access Facebook from mobile devices are almost 50 percent more active on Facebook than nonmobile users.

With so many users and such a relatively low developmental cost, opportunities for making money with these apps abound. Finding out what they are and how you can participate in this gigantic and rapidly evolving market is what this book is all about. This chapter will give you an introduction to this exciting new world. It will answer three immediate questions right away:

1. What are apps?
2. What do they mean for you and your business?
3. How can you monetize them?

What Are Apps?

Everyone talks about apps, but what exactly are they? So many people talk so fluently about apps that sometimes it is a little hard to muster the courage to ask what they are and how they can work for you. Well, here's your answer.

App Architecture

App is an abbreviation for *application*, as in application program. For example, your personal computer–based word processor or spreadsheet is an application. The office management software that runs your office and unites your staff is an application. But today, *app* is more than just an abbreviation—the word has taken on a very specific meaning, reflecting the evolution of software over the last half century.

There have been three generations of applications. Although they have evolved over time, all three still exist and are in widespread use. Here's a brief description of how applications have evolved over the years:

■ **First-generation applications:** These are applications that run on individual computers—a personal computer or a corporate mainframe.

- **Second-generation applications:** These applications are networked with multiple simultaneous users using a single application program, such as a corporate database or office billing system. Many of them are still in use today, and their basic architecture is the same as that used to develop them originally—often in the 1960s and often for minicomputers that no longer exist today.
- **Current-generation apps:** Apps, which are referred to as Apps 3.0 if you want to be specific, are an advance on the second-generation applications. There is still a networking component in most cases and, very often, a shared database. Instead of the user interface being implemented in HTML on the Web or in a programming language running on a local computer, the user interface itself has two components. A framework defined by an application program interface (API) is used and reused for many different applications. A very specific piece of code—an app—is inserted into that framework for each task that is needed. The framework provides much more than just an interface; it is able to perform sophisticated tasks that are unique to that framework and its environment. For example, in the case of Facebook, those tasks allow the app to request a list of the current user's friends; in the case of iPhone, the app can find the device's location.

In this book, the word *environment* is used to describe this common framework or other API—the environment into which the app is placed and in which it runs. We use this specific term because other terms specify programming techniques. A framework, for example, can be used to implement either an app or its environment; similarly, an API can implement the environment or the app.

Apps 3.0 represent an evolution of software architecture in many ways. For years, the idea of building reusable application environments and having a task-specific application has been something of a holy grail. What is different in Apps 3.0 is that the environment not only provides the basics of the user interface, but it often contains a very specific set of interface elements such as those for iPhone or Facebook. *Note:* The iPhone environment (technically referred to as the iPhone operating system, or iPhone OS) powers Apple's new iPad. Almost all iPhone apps run on iPad. Unless otherwise noted, references to iPhone apps include iPad apps.

This new architecture is not limited to Facebook and iPhone, but they are the first two major players in this area. Google is now releasing comparable environments, and other companies such as Palm are also positioning themselves for this new future. By looking at the early players this field, this book will help you understand expected developments and how you and your business can become another player in this exciting and growing field.

■ ■ DOING IT ANOTHER WAY

Both Facebook and iPhone provide alternative architectures in addition to apps. Both let third-party developers write software that links deep into the structure of the environment.

On iPhone, Apple calls them *iPhone Web apps*. These are Web pages that you build in exactly the same way you would any other Web page. When the pages are displayed in iPhone's browser, they can access certain features of iPhone directly. In the Facebook world, this alternative architecture is provided by *Facebook Connect*. You build a Web page just as you normally would, but that Web page incorporates some special JavaScript code that lets your Web page visitors log into their Facebook accounts. After that, your special Web page can use that connection to interact with Facebook.

These alternative architectures provide users with similar experiences to those that apps provide, but there are significant differences. Most important, these Web pages are under your control. People navigate to them in the normal way. Users must have a Facebook user name and password, and on iPhone they must be using the built-in Safari browser. There are no downloads or installs of Facebook Connect Web pages or iPhone Web apps other than the normal behavior of Web pages, as you will see in the following section. This can make these alternatives easier for you to develop and for users to use.

■■■ What Do Apps Mean for You and Your Business?

When thinking about Apps 3.0 and how to leverage this technology to either benefit your business or make money directly from applications, you should consider these points:

- **App development can be inexpensive.** Because apps do not have to implement (or reimplement) environmental functions, they are smaller and less expensive to produce. For the user, that translates into a lower cost—sometimes none.
- **Facebook and iPhone are your partners.** On iPhone, third-party developers create apps, and users buy them from Apple's iTunes App Store. On Facebook, apps are free, but users install them in their Facebook account using Facebook itself. For both iPhone and Facebook, part of the user experience includes having directories of third-party apps.
- **Users have to use iPhone or Facebook to use your app.** Because the primary interaction is with the environment, if a user wants to use an app that runs on Facebook, he or she must join Facebook. If the user wants to run an app that runs on iPhone, he or she needs an iPhone. Thus, Facebook and Apple are encouraging third-party apps to increase their own user base.
- **Apps are often easy to use.** Apps are often easier to use than traditional stand-alone applications not only because the environment provides functionality to the apps but also because that functionality is provided in specific ways that users have to learn only once.

For an app developer, this architecture has other implications. A developer's perspective is different than a user's, so there are a few additional points to think about:

- You need to cultivate your relationship with the environment. In the case of Facebook, you need to become a developer, which entails no cost. It does, however, require you to agree to additional terms and conditions beyond those that bind users of Facebook who will not be developing apps. For iPhone and iPad, becoming a developer gives you access to documentation and the App Store, which costs $99 a year.

- On both iPhone and Facebook, your app must be approved by Apple or Facebook, respectively, before it is listed on the site. (Other models being discussed on other platforms omit this step.)
- Your app must be distributed through the iTunes App Store or through Facebook. If you use the alternative architectures (iPhone Web apps or Facebook Connect), you bypass this step.

■ ■ ■ How Can You Monetize Apps?

Where's the monetization? In the old days (a few years ago), the phrase "Where's the money?" was a common theme. Great ideas abounded, but the question was how to pay for them. (Unfortunately, the phrase has also been used all too frequently in the sense of "Where did the money go?") In today's world of Apps 3.0, the question is "Where's the monetization?"

Monetization has traditionally been defined as the process of agreeing that a certain amount of a precious metal will stand for a specific monetary value, but the term is now also used to indicate the more general process of converting any identifiable object or—more often—action into a monetary value. It is more than just another way of saying, "How can we make money from it?" It is the process of figuring out what the components of any project or product are and which of those components can be moneymakers.

Here are the basic ways apps can be monetized:

- **Sell your app directly to users.** iPhone apps are all available through the iTunes App Store, even if they are free. A mechanism allows users to select apps, download them, and automatically install them on an iPhone.

- **Sell your app once.** If you are skilled at developing apps, you can sell your expertise to people who want to develop their own apps. If your skills are in a specific area, such as marketing consumer products or managing complicated projects, that experience can be valuable to someone with developmental expertise and a vague idea for an app in that area.

- **Promote your existing business.** Use your app as an entrée to your existing website or business. Let people browse your goods and even try out your services with an app. Both Facebook and iPhone now support purchases directly from inside your app through Facebook Credits and iPhone's In App Purchase. These purchases are centered on items that can be delivered electronically either as enhanced content to an app, subscriptions to information, or virtual gifts. Real gifts and donations to causes are now available through Facebook Credits, and there is no doubt that these trends will continue and expand.

- **Sell advertising.** Many of the people who are most interested in the world of apps come from the world of advertising. That is where the extraordinarily large numbers of Facebook and iPhone users carry the most weight. As you will see in more detail in Chapter 12, not only are there many users, but there is a lot of information about them. If your app is of interest to people who are interested in quilting, for example, you have a perfect platform for advertisers who want to reach quilters. The cost of developing a quilters' app is much less than the cost of developing a quilters' magazine or television channel (or even a quilters' website). While your advertising revenues might be fairly low, the low cost of development and distribution can let you make much more money than you would in a more traditional environment.

- **Expand your customer base.** Both Facebook and iPhone are at the heart of the social media world in its new incarnations, as well as in more traditional forms such as phone calls and text messages. Having an app can be part of an image update for your organization. Just having a Facebook or iPhone app can set you apart from the competition. It's great to have a map of your business location on a menu in the local diner or even on your website. But with an iPhone app, you can provide directions to your business from the exact spot where the potential customer is at the moment (using the built-in GPS and mapping tools). You have to make certain that your app provides value to Facebook and iPhone users, of course; otherwise, the whole process can backfire. Done right, there is nothing more efficient for opening doors to new customers than the world of apps.

The final point about monetization is one that was stressed previously: the cost of developing an app and of entering into this world is not just relatively reasonable; it is *low*. (How low depends on exactly what you plan to do with the app; you will find more information about the different types of options and costs throughout this book.) The cost of moving into the world of apps is higher for an international enterprise than for a small regional business, but within each business size, the app world represents a much lower cost than many traditional means of promotion. The cost of experimenting with apps is so low and the potential benefits are so great that you can afford to experiment a great deal to find your way to larger profits.

■ ■ ■ How to Make Money with New Technologies

Before continuing with the specific issues of apps, let's consider a few points that apply to new technologies and have done so for a long time—centuries, in fact. The people who succeed with new developments embrace them fully. The temptation is to consider each new technology in terms of its predecessors and, all too often, as a negative implementation of its immediate predecessor. Consider radio, first known in many places as the "wireless" because it did away with the need for telegraph wires. Cars were "horseless carriages." Even computers have seen this type of characterization: desktop computers were so called to distinguish them from their predecessors, which often took up a special room with its own air conditioning and a raised floor under which cables could easily be run. Those antediluvian computers were not called "room-size computers"—they were just computers, and the name of the newfangled personal versions had to carry the distinction. Every time you think of how to do something with an app that you can do in another environment, go through a mental checklist to see if the activity could actually be done differently in what is a very different type of communication and computing environment. Simply moving an existing process into the world of apps is often not the game-changer that is most successful.

You need look no further than spreadsheets to see how this recognition of a new type of device can capture the imagination of throngs of people. The first applications on personal computers were stripped-down versions of mainframe applications or those for minicomputers and even word processors. No one had seen a spreadsheet program before VisiCalc arrived on the first Apple computer. VisiCalc did something new; it wasn't just a word processing application implemented on a personal computer instead of a dedicated or networked word processor.

Likewise, Photoshop created a new way of working. Although designers and graphic artists were accustomed to using airbrushing and other methods to touch up photographs and other images, the move to digital manipulation as exemplified by Photoshop truly turned it into a new operation.

Merely making an older technology faster sometimes transforms it into something new (computer games are a great example of this), and sometimes it just makes the old technology . . . faster. Until someone comes along with a transformative idea.

The pitfalls come when you are trapped in an old way of doing things or in doing old things. Keep your eye focused not on the next big thing, but on the next, next big thing.

▪▪▪ Introduce Yourself

There are people today who have not only used computers all their lives, but used modern computers and software all their lives. Many people can remember the first time they used the Internet or the Web, but for others, these things have always been present.

The most recent upheaval in the Internet world occurred within the last decade, and it is one of the most critical, because it has changed the way in which people interact with the Internet itself and with other people on the Internet. It is the use of real names. Facebook was a major player in this area, coming on the heels of earlier but smaller real-name sites such as LinkedIn.

The problem arose at the beginning of the computer era when handling large strings of text was difficult. If you packed some text into a single computer word (which is the unit of data that is most easily retrieved and stored), it was easy to handle. Depending on the particular computer and the type of character encoding it used, a word could often store six to eight characters. That was the genesis of six-to-eight-letter user names. When you created an account on an early Internet service provider such as America Online, Prodigy, or Compuserve, you were first asked to select an account name, or *handle*. Actual names were usually too long and were not unique, so they made bad account names. This gave rise to an online community where, in the words of a famous Peter Steiner *New Yorker* cartoon from 1993, "On the Internet, nobody knows you're a dog." (The cartoon depicts a dog sitting at a personal computer and saying those words to a dog who is standing next to him.)

People took advantage of the Internet's anonymity, and in some cases, the Internet became both an anonymous environment and a dangerous one. When you create an account on Facebook, you must provide your real name, and that is the name by which people on Facebook identify you. (You can limit the display of personal data in many ways, but your name is always visible.)

At first, many people were leery of using their real names. After all, for years, they had been warned of the dangers of the Internet and that they should not provide identifying information such as their names. But in an environment such as Facebook's, where all names are visible, people start to behave differently. Sites such as Facebook provide a multitude of privacy settings so users can control what is seen about them and who sees it, but knowing that they can be observed and identified by other people—including friends and relatives in the real world—often changes users' behavior. (This is not to say there are no dangers and nefarious doings among identifiable people on Facebook and the Web, but these risks are much lower than they were even five years ago.)

Whether you are building an app for Facebook or iPhone, you are building it for a specific behavioral mind-set—that of individuals who know they are identifiable. In creating a Facebook account, they provide their name and birthday (to screen out underage users), whereas when purchasing an iPhone they give even more identifiable information, including a credit card number.

This is the new world of Apps 3.0. To profit from it, you need to commit yourself to understanding the possibilities of apps, and that means much more than the basics of their technology. Think about how you can use apps and how your customers can use them. Remember the low cost of entry into the app world, and start to think about how you can provide tangible benefits to your customers (and yourself) within that world.

2

A Closer Look at iPhone Apps

THE FIRST STEP IN learning about apps is to actually use them. The second step is to observe other people using them. Watch for mistakes others make, and make a mental note of the moments they smile with satisfaction. Unless you have a sophisticated user research facility close at hand, your research will be focused on yourself and those around you, and this is harder to do than you think.

You have to explore how users react to apps from two different perspectives: as the person who wants to use an app to do something and as the person watching that person struggle with using the app. Whenever you make a mistake while using an app, see if you can figure out why it happened: Did you simply not understand its functionality, or is the procedure for performing a task not intuitive (at least not to you)?

Once you have mastered the ability to observe yourself using an app, you have a third role to play. You need to understand the basic architecture of each app and its environment so you can say, "Yes, this is done by using the Core Data framework," or "This is a result of a programmatic call to the Facebook friends list." Learn to look behind the scenes of the app so you can learn from its successes and failures.

Thus, using your iPhone or Facebook becomes a threefold task:

- **Use the app.** Push its capabilities and try new things. You want to see what the options are and what range of capabilities you have. Then think back on what you have seen and done and consider what's missing. Where is that space where a new app would make it possible to do something that no other app has yet done?
- **Watch yourself using the app.** Look for the failures and successes. This, too, can lead to opportunities. Maybe an app that combines features from two other apps is something you want to explore. Often, you will find a new way of thinking about the tasks you are doing with your iPhone or Facebook.
- **Analyze how the app works.** As you become familiar with the architecture of apps and the environment, you may find better ways of implementing tasks, which can also be your golden opportunity.

In this chapter, you'll learn more about the functionality of iPhone apps, while the next chapter is devoted to those for Facebook. These chapters are not designed to make you a programmer, although this can be a jumping-off point if that is the direction you want to take. They are slightly technical, but once you have finished learning the groundwork, you can get on to the process of making a contribution to the world of apps—and letting the world of apps return the favor. Let's take a closer look at the world of iPhone apps.

How iPhone Apps Work

The various types of iPhone apps work in similar ways. Following are the striking features that users expect to find.

iPHONE APPS VERSUS iPHONE WEB APPS

It's important to understand the distinction between iPhone apps and iPhone Web apps, as there is a tremendous difference between these two:

- *iPhone apps* are written to run on iPhone in the iPhone operating system.
- *iPhone Web apps* are written to run on iPhone in its Safari Web browser.

iPhone apps are developed using Apple's development environment, which includes the tools Xcode and Interface Builder; iPhone Web apps are developed using Apple's Dashcode development tool. Programming for iPhone apps is done using the Apple development frameworks, including Cocoa; the primary language is Objective-C. iPhone Web apps use HTML, CSS, and JavaScript. Despite the enormous differences in their structure and development environments, users often do not notice the differences.

Because both types of apps have access to a number of iPhone functions and features, users may not go much beyond that. The look and feel is very much the same to the casual user, who focuses on the functionality.

There is one big distinction of which users are aware, although they may not realize its significance. iPhone apps are downloaded from the iTunes App Store; whether or not they are free, you still go through the purchase mechanism. If you have downloaded an app from the iTunes App Store, it is an iPhone app.

iPhone Web apps are downloaded as Web pages, albeit very sophisticated Web pages. You do not go through the iTunes App Store.

iPhone has a built-in Web browser—Apple's Safari. This means that you can view Web pages on your iPhone. iPhone Web apps are sophisticated and enhanced Web pages, but it is worth noting how the pages themselves appear on the iPhone. Apple separates Web pages that are not designed as iPhone Web apps into two categories:

- **Compatible with Safari on iPhone.** These Web pages are presented by Safari according to the standards of the Web, which determine how any browser should present them. In accordance with those standards, if part of a page cannot be displayed, the situation is handled properly. The most common parts of Web pages that Safari on iPhone does not display are plug-ins, Java, and Flash. If part of a page cannot be displayed, there is no crash or collapse: the rest of the Web page displays properly, and an icon replaces the Flash animation or whatever is not supported.
- **Optimized for Safari on iPhone.** These Web pages have no unsupported features (Java or Flash, for example). In addition, they may use iPhone features and services such as e-mail and telephony. On other types of devices, these pages simply skip over the iPhone features, and they are normally coded to detect when they are or are not being displayed on an iPhone.

▪ ▪ iPhone Design

Apps don't come with instructions, and that's part of their attraction. They are simple enough that you do not need a user manual, and their capabilities are evident to a user who has minimal experience using iPhone. This is made possible, in part, by a set of design guidelines that Apple has created and shared with third-party developers. Innovation and experimentation in the content of your app is encouraged and welcomed, but developing an alternative interface is not. In fact, in the case of apps submitted to the iTunes App Store (as well as those submitted to Facebook's Application Directory), a lack of adherence to the interface standards and guidelines can be grounds for exclusion.

The design guidelines for iPhone have evolved over time. iPhone apps are clear descendants of a prior Apple technology. The Mac OS X operating system runs Macintosh desktop and laptop computers. Version 10.4 ("Tiger") was released in April 2005. One of its new features was Dashboard. At the touch of a button (by default, the F12 key), the desktop appeared to dim and an overlay appeared in front of it, as shown in Figure 2.1.

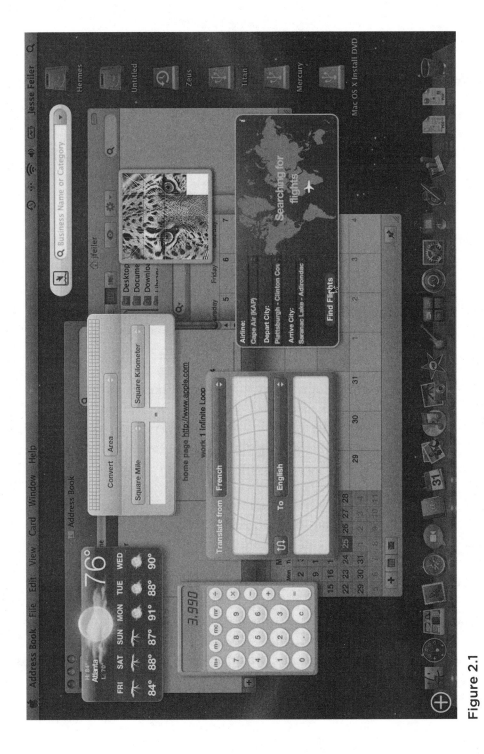

Figure 2.1

Small objects called *widgets* appear on this overlay. As you can see, they focus on a specific functionality. The widgets shown in Figure 2.1 are among those that ship with Mac OS X. From the upper left, in clockwise order, they are a weather widget, a unit converter, a tile game, a flight tracker, a translator, and a calculator. They are different shapes, but they are all small. More important, they lack some of the major components of traditional windows on any graphical user interface:

- There is no drag bar. You can move them around by dragging in any part of the widget except for a data entry field.
- There are no scroll bars. For many widgets, what you see is totally what you get. There are two exceptions to this. Some widgets can flip horizontally. Clicking the upper right of the flight tracker widget will flip it over to display the credits. Flipping the weather widget lets you choose a new location. The icon to flip the widget (usually represented by an italicized letter *i*) appears only when the mouse is over the widget. In addition, some widgets have two displays for their front. For example, the weather widget allows you to click the sun to show or hide the day-by-day forecasts. There is a good deal of flexibility, but there are no scroll bars.

Behind the scenes, there is another critical aspect to widgets: many have access to the Internet. All of the widgets in Figure 2.1, except for the tile game and calculator, have this capability. (Even the unit converter widget has Internet access so it can convert currencies using current exchange rates.)

What you see in Figure 2.1 are Dashboard widgets, but they look and function very much like iPhone Web apps. They use HTML, CSS, and JavaScript in their implementation. Their interfaces are small, designed for the screen of the iPhone or to be placed in the Dashboard layer floating above the desktop. (The screen of the iPhone is technically its *viewport*.)

Apple's user interface guidelines for iPhone Web apps are, not surprisingly, much the same as those for Dashboard widgets. Furthermore, the iPhone Web app guidelines are much the same as those for iPhone apps. The fact that iPhone Web apps and iPhone apps are built using different tools does not really have much to do with what users expect from the interface.

Know Your Users and What They Want. This is a critical design consideration for everything that is well developed. If you are creating a tool that will help people find the latest information about an agricultural pest or disease, what you come up with should look very different depending on whether it is designed for farmers, consumers, commodity brokers, or government agencies. iPhone makes this more important than most other environments because of its size: you do not have room on the viewport for a Welcome From Our Company message, a Join Our Mailing List button, and—by the way—some pest and disease information.

The small viewport means you have to be extremely selective in your content and its presentation. If several people are involved in the development process, try to get them to behave in the opposite way from how most committees are accused of operating. Instead of coming up with a final work product that contains a bit of input from everyone and is therefore big and unwieldy, see if you can get your group to focus on what it can remove. The mantra should not be "Good idea; let's add it," but rather "Good idea, but we don't have room."

You can save space and focus your material by knowing your users and defining them specifically. Because of the vast number of iPhone users, you can target a fraction of a fraction of a small percent of them for your app and still have a large audience. (This is true of Facebook as well.) In many cases, the more specific your target audience is, the more efficient your communication can be.

Put the Browser in Its Place. In order of complexity (for the developer), four types of content can be provided on the iPhone:

- *Compatible with Safari on iPhone*, which includes the bulk of Web pages that have not been modified at all for the iPhone. They are shown in Safari on iPhone.
- *Optimized for Safari on iPhone* Web pages have iPhone accommodations and enhancements. They are shown in Safari on iPhone.
- *iPhone Web apps* go beyond the optimized pages for Safari. They are designed to fit the viewport without manipulation such as zooming or panning. They may include their own navigation tools rather than those for Safari.
- *iPhone apps* run on the iPhone operating system (OS) and may display Web data. Although the WebKit can retrieve and format the data (as it does for Safari), Safari does not manipulate the Web content.

As you can see, the Safari Web browser is critical in the first three types of content—at least from a technical point of view. From the user's point of view, minimizing the role of Safari in any of these types of content makes it more iPhone-like. iPhone Web apps and optimized for Safari on iPhone pages can both be designed to fit exactly into the viewport. That alone makes them look more like an iPhone app than a Web page.

Optimize the iPhone Experience. This means optimizing what you are doing as well as what users may want to do with their iPhones. As computer resources have grown faster, more powerful, and less expensive, developers have used them aggressively and, some would say, without too much concern for the consequences. This is an old story, and it is not specific to computers. Scarce and expensive resources such as long-distance telephone calls were used with great care fifty years ago. As prices came down, people began to think little about picking up the phone to call, say, London or Brisbane from Los Angeles. When people were building Web pages to run over dial-up connections on the early personal computers, they often hesitated to use anything that might slow down the loading of the page. Graphics, for example, were routinely turned off by dial-up users.

Although the iPhone's connections are generally faster than the dial-up connections of old, you still have to be careful when using them. For one thing, you need to minimize the amount of network resources your page or app requires in order to function, and that is true whether that page or app runs on an iPhone, on a desktop browser, or in an app on another device. The second consideration is perhaps more important on the iPhone than on a desktop computer. The iPhone can use a variety of connections to get to the Internet. It may use a second- or third-generation connection over the telecom network, or it may use one or more WiFi networks. It is designed to make the transition from one connection type to another seamlessly. A user in motion might move from a third-generation connection to a relatively low-speed WiFi connection, then on to a second-generation, and then to a very fast WiFi connection. This unpredictability can be very frustrating, as an app or page suddenly speeds up or slows down for no apparent reason, and your app or page can be blamed.

Consider this scenario. Someone walks to work each morning at about the same time. It is a twenty-minute walk, and the person takes the same route each day, including a stop midway to buy a cup of coffee. The person may check e-mail and some Web news sites during the first part of the walk. The cup of coffee is welcome after the e-mail, and then it is on to work. For the last part of the walk, the person may use your app (an optimized for Safari on iPhone page, an iPhone Web app, or an iPhone app). A reasonable person may well notice that your app is much less responsive than e-mail and Web browsing—much, much less responsive.

In fact, the scenario outlined here fits neatly into a transition from a 3G network (available on the first part of the walk) to a fast WiFi network (in and around the coffee shop) and then to a much slower 2G network on the last segment. But your app gets the blame for the differences.

You can do nothing about the variability of connections, but you can minimize your use of network resources, and that is what you should always do. First of all, review your Web interactions with an eye for anything that is unnecessary. When you are bringing down an image to be displayed on the iPhone, there is no point in using an image that displays perfectly on a large desktop display. It should be the best quality for displaying properly on the viewport. This applies to optimized for Safari on iPhone pages as well as to iPhone Web apps and iPhone apps.

Other optimizations directly involve the interactive user experience. Use interface elements and techniques from the built-in iPhone apps so that what is new in your page or app is its content and functionality, not its interface. This is an area fraught with pitfalls, because if you are building a page or an iPhone app that complements your organization's existing website, there may be tremendous pressure to reuse your site's look and feel. Doing so may bind your iPhone site to your external website, but it can make it much more difficult for people to use.

"Hold the Phone." This idiom from the quaint days before hold buttons and call waiting still resonates. Using a telephone has always required only partial concentration. Whether they are doodling or sipping coffee on an old-fashioned landline phone or checking a calendar, address, website, or another call on an iPhone, users expect to switch among activities. This means that as the creator of a page or app for the iPhone, you must plan for the very real possibility that people using your page or app may move on to another page, app, or activity at any time. Whether or not they will ever return (in a brief moment or months later) is a question even they can probably not answer for certain. Every nanosecond that someone spends looking at your page or app may be the last nanosecond he or she spends with it. Having to log out or save work is not a part of many iPhone experiences.

The reverse is also true. You cannot tell which moment will be the last moment someone spends with your page or app, and you cannot tell what will be the first moment and the first page someone will see. Although there are exceptions to this, by and large, people need to be able to jump into the exact part of your app or page that interests them without extraneous hurdles and obstacles.

This ability to jump into and out of a page or an app has an interesting consequence. If people know that it is fast and easy to get in and out, they may stay longer and become more involved with your page or app than they would with one with more complicated entry and departure points. How often have you used a quick reference page or app on your iPhone to look up something that you think will take a second or two and then discovered that half an hour or more has gone by? Knowing you can leave and quickly come back may encourage you to stay and explore.

Less Is More. Avoid decoration on your pages and apps; there is no room for this on iPhone. (Decoration in this sense means any visual element that does not have an obvious and necessary purpose.) Do not hide functionality. Anything your page or app can do should be evident and visible from the start. "But what," you may say, "do we do with the necessary instructions people must read before using the page or app?" The answer is to rethink your interface—over and over, if necessary. If someone approaches your page or app and must immediately make a choice, see if it is possible to remove the choice. If a choice remains, refine it so it is simple. And if a choice is made, remember that Safari on iPhone supports cookies, so saving the choice is often a good idea. (Don't make it a trap; make it easy for the user to change the settings.)

Finally, when thinking about your own app, use the iPhone features and interface elements wherever possible. This can be difficult if you have spent a great deal of time and money developing your existing website. Those custom-designed buttons do not belong on iPhone (at least without modifications). That nifty rollover navigation controller also is not for iPhone. Even your logo is probably not right. Remember to use relevant graphics from your website on iPhone, but adapt them for the size of the iPhone viewport. The iPhone version of your regular logo may look the same to most people, or it may be a reference to the colors, shapes, style, and content of your logo without duplicating all of the details and nuances.

■ ■ ■ Evaluate Your iPhone Presence

Unless you are planning a start-up company, the first and most urgent aspect of your iPhone project is to evaluate what you already have. If you have a website or use e-mail, you are probably already on iPhone for at least some of your users who access these on their iPhones. You may have to play a bit of catch-up, and it is probably a good idea to do so before moving on. Here are your most urgent needs.

■ ■ Evaluate Your Website on iPhone

One of the first orders of business is to check out your website on iPhone. With millions of users worldwide, chances are that at least a few people (and probably more) are viewing your website this way. If you have not done anything to adapt it, it probably does not fall into the optimized for Safari on iPhone category. That means it is either compatible or worse. This section talks about the most common issues you need to address to make your website presentable, functional, and appealing on iPhone.

Get Compatible with Safari. If your website does not run on Safari, you're in trouble—and not just with the iPhone. There are two general reasons why websites may not run on Safari:

- The site may require features, technologies, or plug-ins that are not available on Safari. Because Safari is standards-compliant, if the site does not run on Safari, it may not run on some other browsers.
- Believe it or not, there are some sites out there that can run on Safari but are prevented from doing so by code that requires another browser. There are cases where such code might have been needed in the past and has simply never been removed.

For reasons of history (and also inertia), certain types of sites and industries seem to have more browser limitations than others.

The financial industry, for example, has a lot of issues with browser compatibility.

Some older sites are nightmares because they have a variety of legacy code on them. Rather than revising such an existing site, you may be better off starting from scratch with a new site specifically geared to either be optimized for Safari on iPhone or be a true iPhone Web app or iPhone app.

Get to Optimized. Users can immediately see that your site isn't optimized for iPhone, even if they don't know the official term. All too often, they will quickly use a much less specific type of terminology, such as "That site doesn't work on iPhone." From your point of view, that can be a serious issue, particularly if the next sentence out of the user's mouth is "Let's find a site that works." The main aspects of optimized for Safari on iPhone are just good site-coding practices. If the site uses plug-ins, it should check for them and have a way of continuing gracefully if it needs to deal with missing plug-ins. Far too often, website developers have taken the easy way out: they simply require a certain browser rather than handling each plug-in as they come to it in the code.

▪ ▪ EVALUATE YOUR E-MAIL ON IPHONE

E-mail is one of the most basic Internet protocols, but it, too, needs to be examined in the context of iPhone. Here are some of the things to check out. As with websites, you are looking at iPhone issues, but most of these issues apply in other contexts as well. The main point to remember is that when you send e-mail to customers, they may be reading it on an iPhone. That means you need to pay some attention to the messages you send no matter where they are sent from.

E-Mail Signatures. The first step is simple: make sure your signature fits on the iPhone viewport. Does it contain a graphic that can be optimized to download faster? Is the graphic too wide or tall for the viewport?

The second step is to make certain that your signature contains your contact information as clearly as possible. You may not have noticed what the iPhone is doing to your e-mail text because it happens so elegantly and easily. iPhone is very good at picking out telephone numbers and addresses, then converting them to links in your e-mail messages. Tap a phone number in an e-mail message, and iPhone will ask you if you want to call it. Tap an address, and it is automatically mapped for you. Of course, Web addresses respond to a tap by opening in Safari on iPhone. Make certain that your signature is recognized by iPhone. For example, "One Hundred Fort Scott Drive" may be a correct address in a rather elegant format, but it will not be a link to a map. Likewise, "Arlington, Virginia" is a correct and formal address, but it may not link to a map. You may think that two-character state or province abbreviations are awkward, but the fact is that "Arlington, VA" will produce a link to a map, whereas "Arlington, Virginia" will not.

This recognition of phone numbers and addresses applies not just to e-mail signatures, but to all text in e-mail messages. Mail on Mac OS X performs the same recognition along with sophisticated recognition of dates and times.

Attachments. iPhone displays a wide array of attachment files: Word documents, Excel spreadsheets, documents from Apple's iWork suite, text files, PDF files, and many more. If you routinely send or receive e-mail attachments, check that the formats are compatible with those for the iPhone. This is much easier than it may first appear because there are so many compatible formats.

■ ■ MANAGE YOUR OTHER WEB RESOURCES FOR IPHONE

If you rely on other Web resources, check that they are compatible with iPhone. Calendars, for example, can be problematic. iPhone integrates and synchronizes with many calendar environments, but some people use idiosyncratic or older calendars that do not work with iPhone.

■ ■ ■ How iPhone Web Apps Are Built

As noted earlier, Apple released the specifications for Dashboard widgets to third-party developers when Dashboard was released in Mac OS X 10.4. Soon a wide variety of widgets began to emerge. With the release of the next version of Mac OS X (10.5, or "Leopard"), a new developer tool, Dashcode, allowed third parties to create Dashboard widgets more easily. When the iPhone Software Development Kit (SDK) is installed, Dashcode can help people easily create iPhone Web apps, not only because it is a powerful tool, but also because it ships with a variety of templates for iPhone Web apps. As a result, you can build an app to browse data such as your catalog by clicking to create a Dashcode browser project and importing your data in a file: the entire interface is built for you, and your data is placed appropriately.

Dashcode hides the code from you (something some hand-coders do not like). Creating a basic iPhone Web app often involves little more than filling in the blanks in Dashcode, as shown in Figure 2.2.

Dashcode comes with a variety of templates that just require some customization to work. You can add to the necessary customization to include additional functionality. Even with no further customization, the iPhone Web app that you create is an incredibly inexpensive tool to explore your ideas.

A manager or an entrepreneur can use Dashcode to demonstrate what he or she wants in a final iPhone Web app. For example, one of the templates lets you build an iPhone Web app that displays an RSS feed. Once you select the template, you see the window shown in Figure 2.2. Although there is a lot in the window, the components always appear in the same place, and there are lots of tools to assist you.

In the lower left corner is a scrolling list of workflow steps. These are the steps you have to perform to build the template. Some are required and are marked as such. Each step can be marked as Done or Not Done. Furthermore, within each step, you will find one or more arrows to data entry fields. For example, clicking the RSS Properties link in the Provide RSS Feed step will take you to the field in which

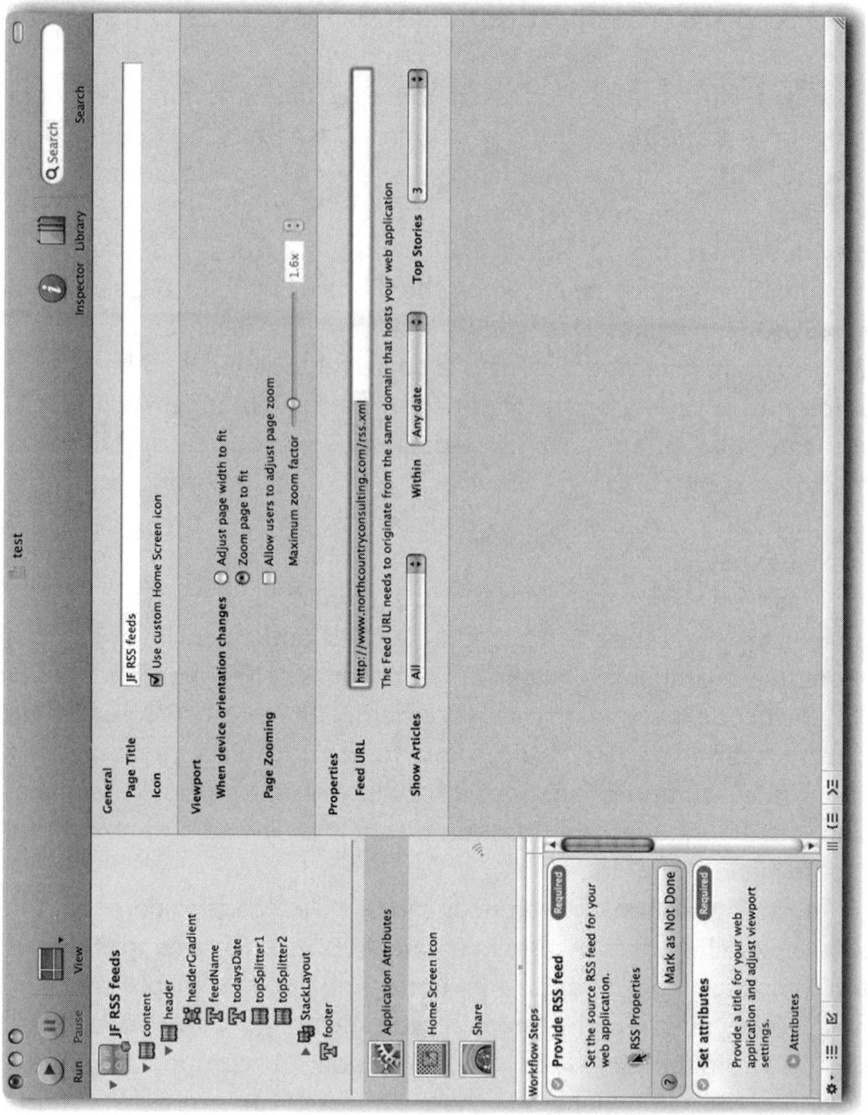
test
Run
Pause
View
Inspector
Library
Search
JF RSS feeds
content
header
headerGradient
feedName
todaysDate
topSplitter1
topSplitter2
StackLayout
footer
Application Attributes
Home Screen Icon
Share
Workflow Steps
Provide RSS feed
Required
Set the source RSS feed for your web application.
RSS Properties
Mark as Not Done
Set attributes
Provide a title for your web application and adjust viewport settings.
Attributes
General
Page Title
JF RSS feeds
Icon
Use custom Home Screen icon
Viewport
When device orientation changes
Adjust page width to fit
Zoom page to fit
Page Zooming
Allow users to adjust page zoom
Maximum zoom factor
1.6x
Properties
Feed URL
http://www.northcountryconsulting.com/rss.xml
The Feed URL needs to originate from the same domain that hosts your web application
Show Articles
All
Within
Any date
Top Stories
3

Figure 2.2

you specify the address of the RSS feed you want to display. Just type it in and mark the step Done.

Above the workflow steps is a series of three editors. These are forms with fields to be filled in or, in the case of the Home Screen Icon, a field into which you can paste an image. You can switch among these forms just by clicking them. The Application Attributes form is the one you see in Figure 2.2.

Finally, in the top left-hand section of the window is the tool to select parts of the interface such as buttons. These are actually drawn in the *canvas* at the right of the window.

Figure 2.3 shows how you can work on the default interface. The elements are shown in a list in the upper left of the Dashcode window. A graphical representation is shown in the canvas at the center of the window. You can double-click anything to change its value. For example, the default title is My RSS Feed, but here it has been changed to Up-to-the-Minute. Just click and type. When you are finished, you can move the files to the Web for testing.

Figure 2.4 shows the iPhone Web app in action. It incorporates the changed title and has retrieved RSS feed items from the URL that was specified. (In fact, that URL is the only piece of data that you absolutely must enter in order to create this iPhone Web app.) As you can see, this feed has entries from a blog; one of the entries contains an MP3 audio file of a radio interview.

You can navigate to the iPhone Web app in a browser including Safari on your iPhone. Wait a while and return to that location; if the RSS feed has been updated, you will see the new items. If the links are all working correctly, you can go to the full entry for each item. In the case of a link that contains additional information (such as the MP3 file), the iPhone does the right thing—in this case, playing the interview.

If you use Dashcode, there is no code to be written. If you want to enhance your iPhone Web app, you can add advertising at the bottom, but that's just a matter of adding a Web element, not writing code. Pick the right RSS feed (or construct your own from a special combination of feeds that your users will appreciate) and you have a valuable resource for users and the advertisers who want to reach

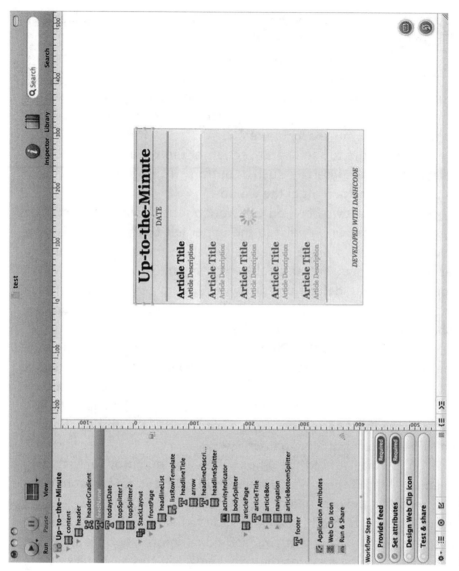

Figure 2.3

Figure 2.4

them. Alternatively, use an iPhone Web app to promote your own content, and it becomes a promotional tool for your own website (or another one).

The point is that the Dashcode template route is fast and inexpensive. You can try out your ideas quickly if you use a template as a prototyping tool. With a little polishing, it can help you create your final iPhone Web app.

■■■ How iPhone Apps Are Built

The power of Dashcode lies in its templates and powerful interface. Xcode, the tool you use to develop iPhone apps, also has these two features. But Xcode has much more, as you can see from some of the features shown in Figure 2.5.

Xcode—and thus, iPhone app development—is for programmers. The templates generate code that you can then extend and customize, but for the most part, you will be working in Objective-C code, a dynamic, object-oriented language. A companion tool, Interface Builder, is shown in Figure 2.6. (Xcode was originally called Project Builder, so the names were symmetrical.) This tool features a graphical user interface with some of the same types of icons and fill-in forms that you saw in Dashcode. Xcode can handle the largest development projects. Your iPhone app is certainly important, but Xcode is also the tool that engineers at Apple used to build the Mac OS X operating system, as well as a wide variety of tools and application programs.

■■■ How to Create an iPhone App

Creating an app is only the first step: you then have to distribute it and get users to use it. iPhone apps are downloaded from the iTunes App Store; once downloaded, they are installed. The steps involved in getting an app to the App Store are as follows:

1. You register as a developer ($99 per year in the United States; developer.apple.com).
2. You develop and test the app.
3. You submit your app to Apple.
4. If it is accepted, it is available through the App Store.

There is no submission process for iPhone Web apps or for Web pages designed for Safari on iPhone. That is one of the reasons so many people create them.

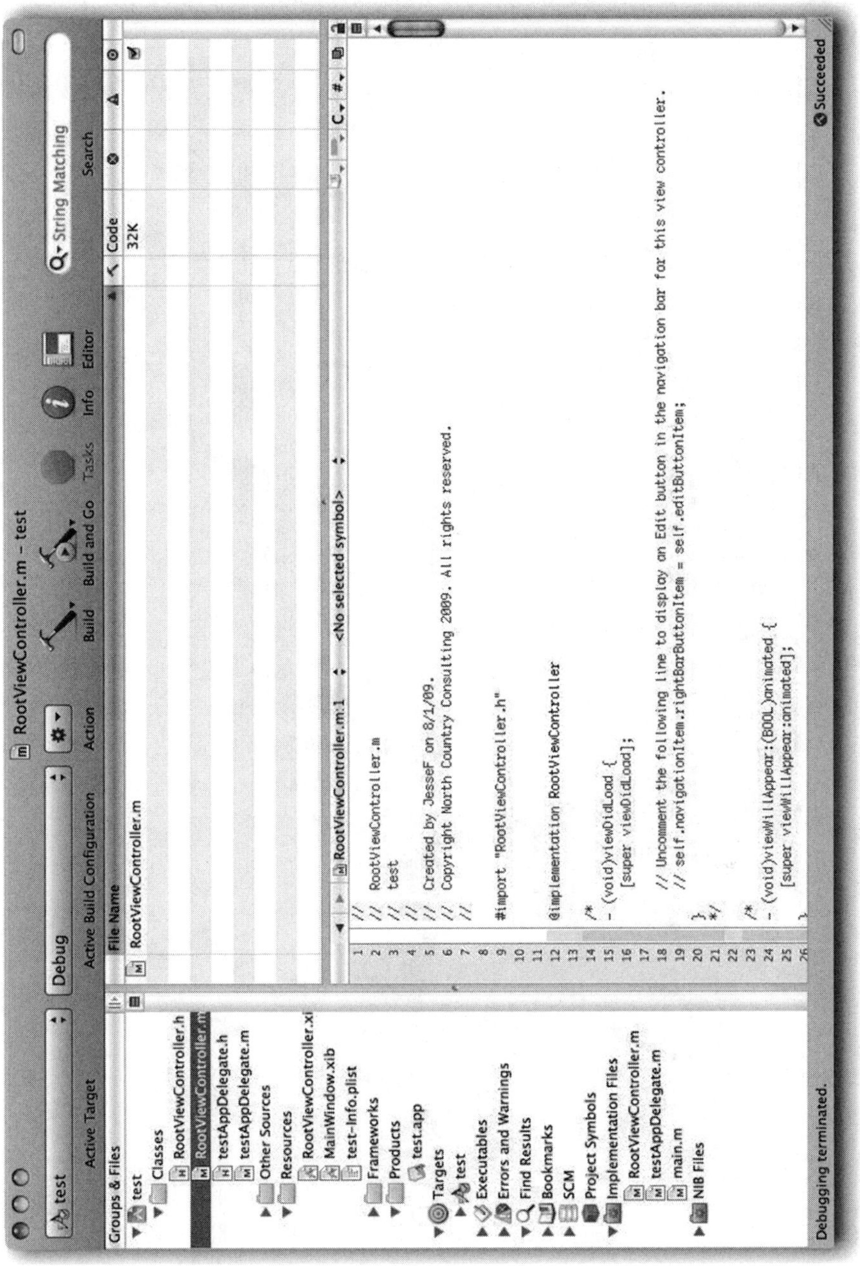

RootViewController.m - test

test   Debug   RootViewController.m

Active Target   Active Build Configuration   Action   Build   Build and Go   Tasks   Info   Editor

Q~ String Matching

Search

Groups & Files

File Name   ◄ ◄ | Code

▼ test
  ▼ Classes
    RootViewController.h
    RootViewController.m
    testAppDelegate.h
    testAppDelegate.m
  ▼ Other Sources
  ▼ Resources
    RootViewController.xi
    MainWindow.xib
    test-Info.plist
  ▶ Frameworks
  ▶ Products
    test.app
  ▼ Targets
    test
  ▶ Executables
  ▶ Errors and Warnings
  ▶ Find Results
  ▶ Bookmarks
  ▶ SCM
  ▼ Project Symbols
  ▼ Implementation Files
    RootViewController.m
    testAppDelegate.m
    main.m
  ▶ NIB Files

RootViewController.m   RootViewController.m:1   <No selected symbol>

```
 1   //
 2   //  RootViewController.m
 3   //  test
 4   //
 5   //  Created by JesseF on 8/1/09.
 6   //  Copyright North Country Consulting 2009. All rights reserved.
 7   //
 8
 9   #import "RootViewController.h"
10
11
12   @implementation RootViewController
13
14   /*
15   - (void)viewDidLoad {
16       [super viewDidLoad];
17
18       // Uncomment the following line to display on Edit button in the navigation bar for this view controller.
19       // self.navigationItem.rightBarButtonItem = self.editButtonItem;
20   }
21   */
22
23   /*
24   - (void)viewWillAppear:(BOOL)animated {
25       [super viewWillAppear:animated];
26
```

Debugging terminated.   ✔ Succeeded

Figure 2.5

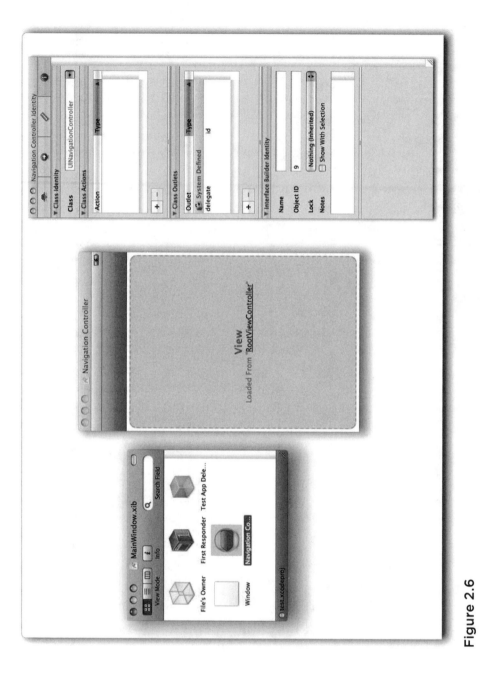

Figure 2.6

■ ■ ■ Back to the Top: Varieties of iPhone Apps

It is time to return to the initial theme of this chapter: the anatomy of an iPhone app. You have seen the anatomy from various perspectives, including those of a user and a developer. Each of the different types of iPhone content has its own development process and its own structure. But what is important to you and to other users is that all of them should look and behave the same—this is particularly true for iPhone Web apps and iPhone apps. In choosing what to implement and how to implement it, you will need to trade off the costs and benefits.

Above all, the focus in your iPhone apps (and in your Facebook apps, too) should be on the functionality. Whether that functionality is a reference to esoteric equations, an engrossing game, or a way of keeping up with friends, that's what is important.

Figure 2.7 shows an example of an iPhone app that addresses a simple issue: finding and contacting a library in the Clinton Essex

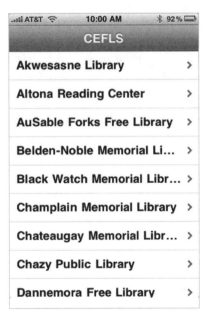

Figure 2.7

Franklin Library System (CEFLS). Using his or her finger, a user can quickly flip through the list of libraries (shown on the left) and then choose one by tapping on it. The screen then displays the contact information for that library (shown on the right).

Built-in iPhone functionality lies behind each icon on the contact information screen. Tap the phone icon, and iPhone calls the library. Tap the e-mail button, and the mail app is launched with a message addressed to the library. Tap the folded map, and you'll find a map of the library's location as well as directions from exactly where you are at the moment to the library. All of this functionality is built into iPhone; what you provide is the idea of building the directory and the specific data that's needed.

The app shown in Figure 2.7 happens to be an iPhone Web app, but it's hard to tell which variety of iPhone app it is by looking at it or using it. You may be most interested to know, however, that the time it took to build the app using Dashcode was just over an hour. And the time it took to type in the data for the thirty-five libraries in the library system was just under two hours.

As you will see, you may find your role not in developing iPhone apps or iPhone Web apps but in related fields such as advertising and marketing. Whatever you wind up doing, knowing how to evaluate both types of apps and how they can be created is a crucial part of your background. The same is true for Facebook apps, which are covered in the next chapter.

3

A Closer Look at Facebook Apps

AT FIRST GLANCE, FACEBOOK and iPhone seem to be very different technologies. The iPhone is a physical product, whereas Facebook is software. You pay for an iPhone, but Facebook is free. Although Facebook was not the first social networking site, it has become one of the most important—if not *the* most important. And although iPhone was not the first smartphone, its effect on the market and on the technologies involved in mobile telephony has been as influential as that of Facebook in its own sphere.

Despite the differences, there is widespread agreement that Facebook and iPhone are two of the most important stories in the technology and business worlds. What they have in common is large and enthusiastic user bases. People are using technology in a way that was not only unthinkable ten years ago but also impossible. And there is a great overlap between the Facebook and iPhone user bases.

Just as the previous chapter took a closer look at the anatomy of the iPhone, this chapter takes a more detailed look at Facebook. Inside the two technologies, some significant similarities appear. Perhaps the most important is that both are designed for third-party enhancements and add-ons in the form of apps. Both iPhone and Facebook have different types of apps that can be developed by different types of developers, so you can choose what type of app to create as well as what type of resources you want to devote to it. The developer communities for both technologies number in the tens of thousands. Most important, the third-party apps bring added value to the technologies and thus to the massive user bases. For the apps' developers and owners, they can provide revenue and other values that quickly can become the core of a new business or a significant addition to an existing one.

Understanding how Facebook apps work requires that you use the same process you did to understand how iPhone apps work:

- Use Facebook.
- Watch yourself using Facebook.
- Analyze how Facebook functions.

■■■Facebook from an App's Point of View

Most of the things users can do on Facebook are very simple. They ask someone to be a friend or respond to a friend request. They can join a group or create one. With a mouse click or two, they can share photos, videos, comments, or URLs they have found on the Web with their friends. Figure 3.1 shows the types of interactions users can have with a Facebook app; they are common processes. This app lets users browse sports teams in a specific geographic area and select those that interest them. None of this is new; you have been able to do things like this on the Web for years.

What Facebook does is turn this into a social network, and that enables the swift spread of information that is a hallmark of the social networking world—an opportunity to get more exposure for your own app. The first aspect of this is visible if you look closely at Figure 3.1. In addition to being able to browse the teams and view your own list, you can see a list of your friends' teams. This is not the list of a specific friend; rather, it is a list of all the teams for all your friends

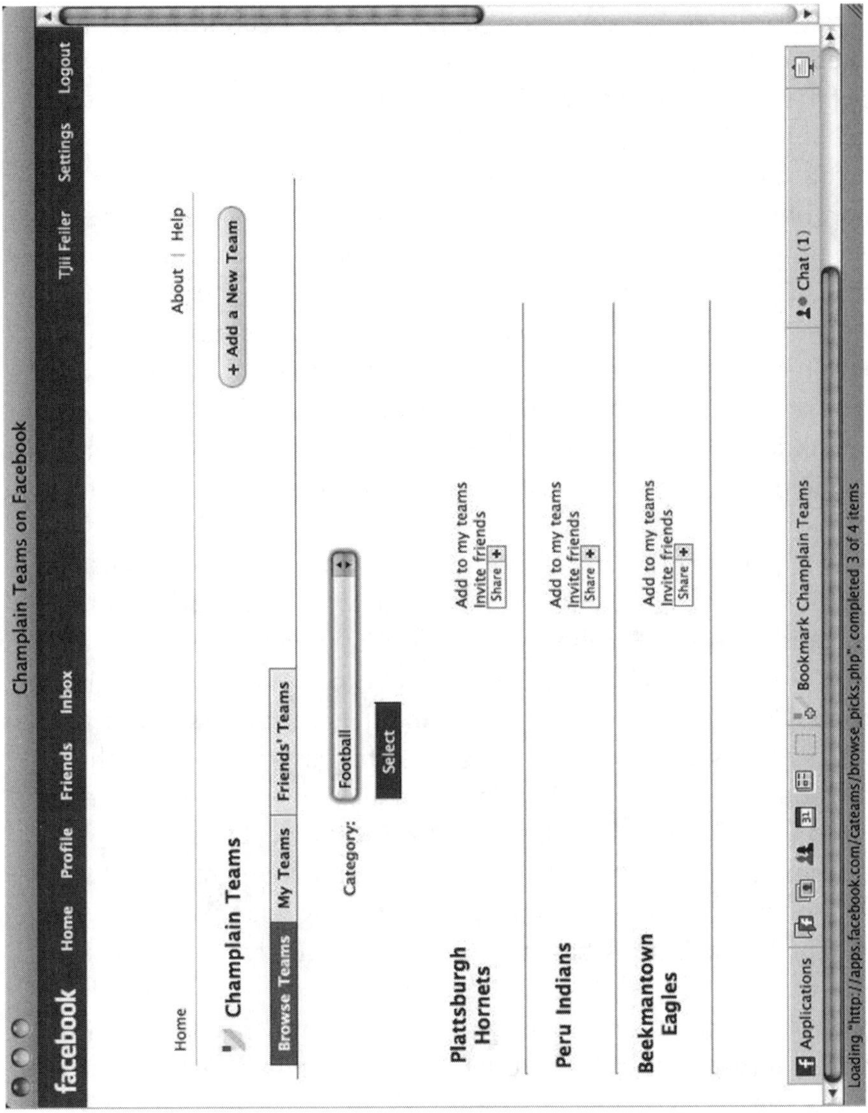
Champlain Teams on Facebook
facebook
Home
Profile
Friends
Inbox
Tjil Feller
Settings
Logout
Home
About | Help
Champlain Teams
+ Add a New Team
Browse Teams
My Teams
Friends' Teams
Category:
Football
Select
Plattsburgh Hornets
Add to my teams
Invite friends
Share
Peru Indians
Add to my teams
Invite friends
Share
Beekmantown Eagles
Add to my teams
Invite friends
Share
Applications
Bookmark Champlain Teams
Chat (1)
Loading "http://apps.facebook.com/cateams/browse_picks.php", completed 3 of 4 items

Figure 3.1

merged together. Because Facebook friends are confirmed by both parties, and also because they are real people with real names, this list of friends and their teams has value and authenticity. That is what Facebook's friend structure provides.

Still, this is a basic and somewhat old-fashioned type of application until you look at a user's news feed, as shown in Figure 3.2. Notice that the first item shows the user has selected Peru Girls Basketball. This item appears in this user's news feed, but it also may appear in the news feeds of that user's friends. The item contains a link to the team; in this case, the link goes to the team's entry in the app's database. That item can contain additional information as well as a link to the team's own website. The user can then comment on these items as well as indicate whether he or she likes the item or not.

This item is automatically generated by an app when a user takes an action (selecting a team). The app structures this information and then sends it off to Facebook. The information appears in the news feed of the user who is using the app on his or her profile page. Then Facebook's software gets to work behind the scenes. First it applies a host of security settings: what does the acting user allow to be shown to friends? Then a series of preferences kick in: what do the friends want to see? (Those Like votes can help Facebook at this stage.)

Because news stories generated by an app are highly structured, Facebook can combine them. If you are viewing your own news feed and your friend User A has picked a team, you may see that news item. If another friend, User B, has done the same thing, Facebook can combine those items to tell you, "User A and User B picked Peru Girls Basketball." User A and User B may not know each other, but because they are both friends of yours, their news may appear in your news feed, and Facebook can combine them in this way. This creation of a story based on the actions of your friends is a powerful Facebook tool: you can watch an app or a business or an idea spread across your circle of Facebook friends—and their friends.

The news item retains any links that the app may have created. The idea behind all of this is that your news feed is customized for users and their preferences; it includes news items about their

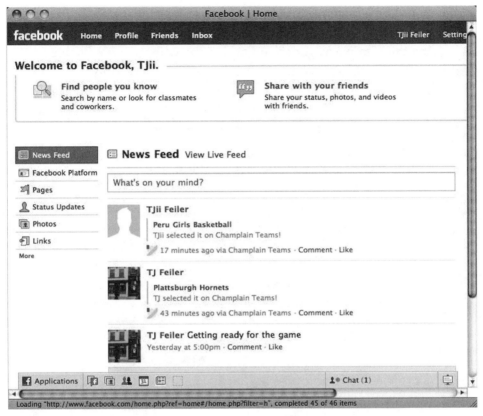

Figure 3.2

friends; and individual stories can be generated by Facebook through the combination of separate events. This tends to make entries in a news feed more interesting, and if they are more interesting, people are more likely to click on those links.

Also, if you see that many of your friends are trying out a new game or reading the same book, you may take that as an implicit suggestion that you should join in. (Facebook has a formal invitation mechanism by which friends can explicitly make such suggestions.)

The second item in this news feed has also been generated automatically by an app: it reflects the selection of the Plattsburgh Hornets by another user. (These are test accounts with very similar names; they are available through the Developer App on Facebook.

You should not attempt to gain access to them or to make friends with them, because your own account can be compromised.)

The third item was generated by a friend who has typed in a comment ("Getting ready for the game") in the "What's on your mind" box at the top of the news feed. Thus, the news feed is built automatically based on Facebook rules and app code, as well as by explicit user and friend updates. It is one of the most interesting aspects of Facebook to many people.

A large part of app development goes to generating these news feed items. The process goes through several steps:

1. Users need to use your app.
2. The app generates news items. This means there are concrete and specific actions that users can take, such as selecting a favorite, recommending a cause, or adding a comment, and the app generates the appropriate news items. (The actions also have to be concise in their descriptions.)
3. The news story shows up in friends' news feeds based on preferences and other settings. There are usually more news items than will fit in the news feed, so Facebook is selective. The more friends who do the same thing, the more likely it is that an item will show up in your news feed.
4. Links encourage users to use the app based either on news stories or on specific invitations from friends. The more people who use the app, the more likely it is that several of your friends will use it and that stories from the app will appear in your news feed.
5. The process repeats over and over.

This is an example of the social network in action. The viruslike behavior as use of an app spreads is called *virality*. This is Facebook from the app's point of view. From the user's point of view, it may be a way to keep up with friends or a way to share things. As you continue to use Facebook yourself, that is really all you will care about.

Facebook provides a number of *integration points* at which an app can interact with a user and the user's friends. The news feed is one of the most important; others are the wall where users and their friends can share messages, as well as user profiles where users can present their own information and display information from apps in which they are interested.

■■■ Aesthetics of Facebook

Facebook runs as a Web application with a recognizable look and feel. Each user has a *home page*, such as the one shown in Figure 3.2. Other pages are used for other purposes; Figure 3.1 shows a typical app page. The basic look of Facebook is fairly stable, but there have been changes over time, particularly within certain areas such as the user profiles.

As a user, you are probably used to Facebook and its navigation tools. Now, as someone who is interested in Facebook apps, you need to look beneath the functionality and the content. Figure 3.3 shows a *profile page*. Compare it to Figures 3.1 and 3.2 to find the common elements.

All Facebook pages are presented within a *frame*, which is visible at the top. At the bottom of most Facebook pages are standard links, and advertisements or announcements often appear on the right. App developers and users cannot control these elements: the frame, the ads and announcements, and the links are provided by Facebook and are under its control. (Facebook selects ads for individual users based on their preferences and behaviors, so in a roundabout way, users influence the ads that appear on their pages but do not control them.)

The center of the Facebook page for apps is the *canvas*, which is under the control of the app. From Facebook's point of view, an app has only one purpose: it must draw on the canvas. That's where you provide value for your users. Because the frame contains all the basic Facebook tools, you do not have to reimplement them (nor should you). The entire canvas is yours to use totally for your app and its functionality.

Remember, as Facebook evolves, so do its features and layouts.

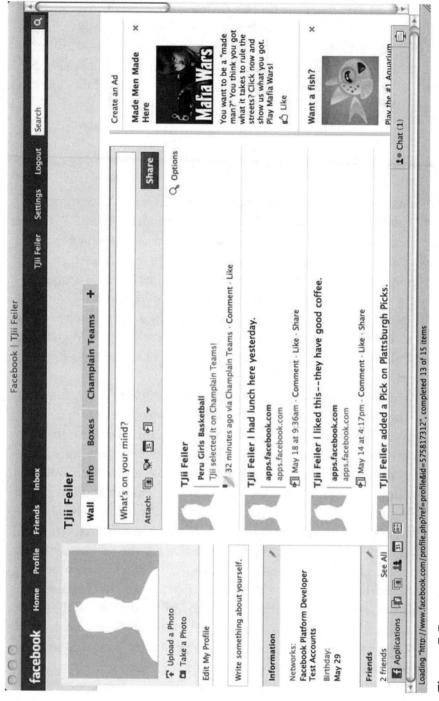

Figure 3.3

■ ■ ■ How to Create a Facebook App

Creating an app is only the first step: you then have to distribute it and get users to use it. You saw the process for an iPhone app in the previous chapter.

The Facebook app structure is somewhat similar and somewhat different. There are only Facebook apps—no variations as there are with iPhone Web apps or Web pages for Safari on iPhone. (You will find out more about Facebook Extensions later in this chapter.) Here is an overview of the process for Facebook app development:

1. Register as a developer by downloading and installing the Facebook Developer App from facebook.com/developers (if you are not logged into Facebook, you will be prompted to do so). There is no charge for this.

2. From inside the Developer App, create a new app. You provide information such as the name of the app and then receive two *keys* for the application. These long strings of letters and numbers will be used in your code. Your app will run on a Web server that you control or to which you have access. You must specify the address of that server. (Technically, what you provide is the location of a folder or directory on the server. You place your app in that folder or directory.) There is no charge for creating an app.

3. You develop and test the app.

4. The app can be used at this point, so you can stop there if you want. However, most people continue on to submit the app to the Facebook Application Directory. Once accepted, the app is listed, and users can search for and download it.

For both iPhone apps and Facebook apps that are listed in its Application Directory, there is a gatekeeper function provided by either Apple or Facebook. You do not need to go through the gatekeeper function for iPhone Web apps and Web pages, nor do you need to go through it for apps that will not be listed in the Application

Directory. Your app will receive increased visibility in the iTunes App Store and the Facebook Application Directory. You need to consider this against a certain additional amount of time for the approval process. Ultimately, your choice will be determined by what you are trying to do. If you are building an app that will be used by your existing customers (perhaps to browse your catalog), that is different from an app that you want to be used by people with whom you have no relationship. A simple rule of thumb is that if your potential users are already using your website, you can probably bypass the Facebook Application Directory and the iTunes App Store. But if they are out there—somewhere—and you have no contact with them (yet), you probably want the broader exposure of being listed.

■■■ How Facebook Apps Work

As described previously, in practical terms, the purpose of a Facebook app is to draw on the canvas within the Facebook frame. Here is how that happens.

The code for your app runs on a Web server that you control or to which you have access. (You specified that address as part of the Facebook app registration process.) Let's say you have called your app *jfbirds*. A user accesses the app from a link or types in an address using a URL of the form http://apps.facebook.com/jfbirds. Facebook can see that this is a request for an app, and that its name is *jfbirds*. It looks up the information for that app to locate the address of the server on which the app runs. It requests that server to send the information for the canvas from the requested page of the app. The code that runs on your Web server returns the canvas data for the requested page to Facebook. As a safeguard, your app also returns the keys from the initial app creation so Facebook can make certain the response is legitimate. You need a server to run your app, and it must be connected to the Internet so it can communicate with Facebook. As long as you provide the server address in your Facebook app registration, Facebook can find it. It does not have to be the same server you use for your website, and in many cases, people prefer to keep their Web server separate from their Facebook server.

When the canvas is returned to Facebook, it constructs the complete page for downloading to the user's browser: the frame, ads, and footer links are inserted as needed.

While your app is constructing the canvas, it may need information from Facebook (a list of the user's friends, for example). Your app, running on your server, can send a message (with the proper key) to Facebook to retrieve the necessary data. In addition to retrieving data, your app may be sending data, such as items for the news feed. Constructing the canvas for the app's page may require a number of similar calls back and forth to Facebook. As you can imagine, the Facebook servers are highly optimized, and the code supporting these calls is similarly created to be as efficient as possible.

The days when a Web page was a single page of HTML code are long past. It is true that some such pages still exist, but more often the "page" either includes other components such as JavaScript and Cascading Style Sheets or is some code that dynamically creates the page. This is true of Web pages, iPhone Web apps, and Facebook apps. (iPhone apps may include HTML, JavaScript, and Cascading Style Sheets, but they also include iPhone and Cocoa frameworks that are usually written in the Objective-C language.)

These are not strange technologies for the most part. For instance, JavaScript is the scripting language that helps autocomplete text as you start to fill in a form on the Web. Cascading Style Sheets are commonly used to separate the formatting of a page from its content— you choose how specific types of data such as a header or a bulleted list will be formatted, regardless of the text they contain. Objective-C is used primarily for programming Mac OS X applications (and the operating system itself), along with iPhone apps.

The most common tools used to assemble an app's canvas are HTML, JavaScript, Cascading Style Sheets, and a programming or scripting language such as PHP. In addition, because Facebook apps often store data, they may use a database—most frequently, MySQL. Although other languages are in widespread use, the PHP/MySQL combination is used much more extensively, including in Facebook. Thus, the tools needed for developing Facebook apps are not specific to Facebook and are not unfamiliar to today's Web designers.

In addition to the tools used, Facebook apps, iPhone apps, and iPhone Web apps share another development feature: they all use rich and robust frameworks of code to make up their environment. These reusable code structures implement the bulk of the functionality for all apps. You do not have to rewrite or redesign most of the environmental functionality. In iPhone and Cocoa, these are true object-oriented frameworks; in Facebook, they are less object-oriented, but the basics are the same. (Object-oriented design is the way most modern software is created. It lends itself to fast and modular development and maintenance.) Most of the code that you need to implement an app for Facebook or iPhone already exists; it is designed for reuse and is easily accessible. (In fact, not using the framework code is almost always a bad idea.)

▪▪▪ Facebook Extensions

A number of extensions to Facebook build on its architecture. One of the most important at the moment is Facebook Connect, which allows you to create a Web page on your own site that lets people log on to Facebook. They provide their Facebook log-in information on your site, and you transmit it to Facebook along with the information you obtained from the Developer App when you registered your app. You are then able to call some of the same functions that you can call from the PHP code in your Facebook app from your website. The processes are the same, and you must always provide the registration information. This enables you to retrieve certain information from Facebook for the user and, more important, to submit stories for the user's news feed based on the user's actions in your website (subject, of course, to the usual options and security constraints).

The other major extension to Facebook is its iPhone app. Facebook's development trends increasingly allow its basic functionality to be accessed by third parties through apps running inside Facebook itself as well as apps running in all sorts of environments. Opportunities are vast as Facebook evolves. The core of Facebook—things such as the network of friends and the news feed—is available in many ways. You should be able to find a comfortable niche in these various implementations to allow you to profit from its success as more and more people use your app.

4

Taking Advantage of iPhone and Facebook Features

IT IS THE FEATURE sets on iPhone and Facebook that attract people to apps like yours. If you want to make money with apps, deliver your own special content and functionality alongside the features that have brought users to the environment in the first place.

To successfully build and promote an app in this environment, you have to observe people—including yourself—as they use the environment. Furthermore, you have to think about what is available to you in Facebook or iPhone so that you can make your app successful.

If you have not experimented with the iPhone GPS system and built-in maps, get started now. Many people find it addicting to be able not only to map a specific location but also to show a route and estimated travel time from where they happen to be at that moment to their destination. And tapping the bus icon often gives public transportation options. Some systems provide enough information to Google and iPhone to let users see the time of the next one or two buses, subways, or trains on their route. Becoming part of this world places your business in the middle of a rich, user-centric environment.

On Facebook, the social graph of friends is similarly fascinating (*social graph* is the technical term for the relationships among friends on Facebook or any social network). Watch as the use of an app or purchase of an item spreads through your friends (or as negative comments abut a business or movie spread just as fast). Your app can be part of this web of friendship, and as you will see, you can guide your app's users to share positive experiences with their friends.

This chapter addresses how apps function in the iPhone and Facebook environments, whereas the next chapter addresses what the apps look like. These are related issues, and they arise from the fact that your app is not running on a personal computer somewhere. It is running inside a Facebook frame that is inside a browser; on iPhone, it may be running inside Safari on iPhone or on the iPhone operating system (OS). In all of these circumstances, features and functions are available that are not available on a desktop. In addition, some features and functions that are standard on a personal computer are nowhere to be found. If you design or build an app for the context of a personal computer, it will fail as a Facebook or iPhone app. You have to forget a lot of what you know about personal computers.

It is important to think about these issues, because when a desktop app is ported to iPhone or Facebook, its failure may not be obvious. It is not so much a matter of an app crashing, but rather, a problem when apps bring functionality from the desktop that does not belong in an app. For example, your entire structure of menus on a desktop app probably needs to be thrown out. Both iPhone and Facebook apps use tabs so you can switch quickly from one view to another, and the notion of elaborate, hierarchical menus does not work, even though you can create such an interface.

Also, the notion of saving your data—which is so essential on a desktop—is much less important to most app users. People expect their mouse clicks and data entry on both Facebook and iPhone apps to work without an extra step to save them. An explicit save command is present in many apps, but it is the exception rather than the rule. As you will see, automatic saving of data makes the app experience better than the desktop experience for many people.

■■■ Using the Code That's Already Written For You

Both Facebook and iPhone apps have an architecture that supports high-level coding. Decades ago, developers had to write almost all the code they wanted to use from scratch. Over time, techniques such as structured programming and object-oriented programming made it possible to share and reuse code.

Today, these architectures provide a high level of functionality that developers need only customize for their own purposes. This chapter is a guide to the functional elements available to developers on Facebook and iPhone—the components from which you build apps. Some of these high-level functions are available in many development environments, and not surprising, a number of them are common to both Facebook and iPhone.

You can divide the elements into three categories:

- *Building blocks* are the basics for building an iPhone or Facebook app. Although important for developers and programmers to deal with, they are not discussed in this book.
- *Common features* are building blocks that support functions such as friends, photos, and media, as well as Web access and e-mail that are common to both Facebook and iPhone. They are the topic of this chapter.
- *Specific features* support functionality such as telephony on iPhone and groups on Facebook. They are described in later chapters of this book.

Don't think that these features are a checklist. The most success-ful apps focus on one or two specific areas, so there is no reason to bloat them with features. However, consider how these features might be incorporated into your app so that functionalities such as location and friends are an integral part of your app. Above all, do not reimplement the functionalities. Users expect to find them in and around their apps, but they expect them to behave in a certain man-ner. Fitting into the environment is one of the key elements of a suc-cessful app.

Now let's go on to the functionalities you get for free with iPhone and Facebook apps. The first two common features are so self-evident that you may not even think about them, but they are very important.

■■■Using the Web in a New Way

An app for Facebook or iPhone/iPod touch runs in an environment in which the Web is available. In the case of Facebook, the apps are served up by a browser with the data from Facebook, so obviously if someone can run your app, he or she already has a browser and an Internet connection. In the case of an iPhone Web app, the user also accesses your app via the Web.

In the case of an iPhone app, there is slightly less certainty (but not much). Because the app is physically located on the iPhone, the user may not have a Web connection at the time the app is being used. This can be because he or she is out of range of a signal (in which case, the lack of a Web connection may be a surprise) or because the user has turned off the Web interface—perhaps by turning on Air-plane mode.

The potential (and, in the case of Facebook apps, almost certain) presence of a Web browser has many consequences. First of all, if your app contains Web addresses as links, to the extent that you control them, those links should be iPhone- or Facebook-friendly so users do not suddenly drop into an abyss when they click on a link.

■■■ Working with Mail

Similarly, your iPhone and Facebook app users exist in an environment that includes e-mail. Both development environments contain tools to let developers give users the ability to send e-mail either directly (iPhone) or through a private messaging service that interacts with e-mail (Facebook).

Incoming e-mail is also part of the picture. You will see how planning for incoming e-mail can make your app more useful to users.

■ ■ SENDING MAIL

Mail is important to your app because its easy availability lets users communicate directly with friends as well as your own company about anything, including your app. Apps spread virally from one user to another, often through recommendations, and those recommendations are often sent in e-mail. It is not just that someone can write a message saying, "I love this new app. You should try it." You can help the spread of your app by letting users interact with iPhone and Facebook's e-mail resources with messages that you prepare and they can modify before they send. Although the option to let users customize the outgoing message is very important, you will likely find that most people do not use the option and instead send your suggested message without any changes.

Before jumping in to generate e-mail in your app, make certain you do your homework. This is one of the areas covered in the terms of service in which your app runs. If an app violates privacy or sends spam, there is a legitimate concern that people will blame the platform rather than the app. As a result, both iPhone and Facebook are vigilant in watching the behavior of apps, and this is one of the most important areas they monitor.

Nevertheless, integrating e-mail effectively with your app can greatly increase its usability and can wind up dramatically increasing the number of users. Whether you are selling your app, selling advertising on your app, or using your app to generate traffic and sales on your website, increasing the number of users is an important first step.

▪ ▪ DEALING WITH INCOMING MAIL

When you think about what it means to be integrated with e-mail, you probably think about being able to send mail. From the standpoint of an app, the ability to receive e-mail is much more important: it can impact the way you design your app.

Mail is the major interruption your users can expect to encounter as they use your app (other than noncomputer interruptions, that is). Incoming e-mail messages and text messages can distract a user. Added to that initial distraction, the user may answer the e-mail or text message and then attempt to return to your app.

This requires that your app be *stateless*: users must be able to leave it or even turn it off and then pick it up or turn it back on without losing anything. The Web itself is stateless, so Facebook—when running on a Web browser—is inherently stateless. iPhone apps are stateless, because when you press the Home button to move to another app, the first app quits. Apps do not continue to run or wait in the background, in most cases.

Programming apps to be stateless means that users can be interacting with a Facebook app, get distracted and move on to something else, and then come back at another time and on another computer without missing anything.

Mostly.

Unsaved information on a Web page is lost when a user navigates to another page. If the user is in the process of entering an appointment in the iPhone Calendar, pressing the Home button ends the Calendar interaction, and the information is lost. There are a number of ways to minimize this risk. The simplest strategy is to avoid lengthy data entry routines. If people are entering data into your app, make the data as focused as possible, and when they move to the next item, store the partial data. This makes the app more useful and intuitive and increases the chances of return visits and recommendations to the user's friends.

In a traditional application, you might ask a user to enter data for a new event. The nicely designed (and large) data entry layout might have space for any or all of the following and more:

- Event name
- Date
- Location
- Notes

If someone forgets to save or navigates away from the Web page once the data is entered, that is the end of the data entry process.

On iPhone, you can split it up easily. (Note that this is a conceptual presentation and not the way the built-in Calendar app works.) Here is what you could do for multipage data entry:

1. Users enter a new event and its name.
2. They click Next, which stores the data behind the scenes, and move on to the date.
3. They enter the date; click Next, which again stores the data behind the scenes; and repeat the process as needed.

Notice that the storing and saving of data is done without direct user action. It is a by-product of the user going to the next screen. Having to go to the next screen to enter more data is a simpler concept to convey than the notion that information needs to be saved. You may think this is self-evident, but the failure to explicitly save data is a common problem for users.

This design strategy also works with Facebook apps. Longtime developers learned long ago to minimize network accesses because of the inherent delays. That gave rise to the notion of collecting a lot of data and then doing one network access to store it all. This design strategy is particularly unsuited to apps that need to be stateless (that is, a large amount of related data elements that all need to be transmitted and updated at the same time).

One way of handling the issue of state is to use another common functional component: data storage.

■ ■ ■ Using the App's Environment to Store Data

On personal computers, many applications use a document-centric data model. When you open a word processing document or a spreadsheet, the entire document is read into memory. You view it and make any necessary changes, and then you save it. Anything that happens between the document load and the document save can be lost if the computer loses power. There are warning messages if you try to close an unsaved document or if you try to quit an application with unsaved documents, but these can be ignored. If there is a problem and the application quits unexpectedly, the changes can also be lost.

Over the years, non-document-centric applications have been developed. They are characterized, from the user's point of view, by the absence of a save command. Internally, they are usually characterized by the use of a database to store data.

Some applications, such as accounting products like Quicken, use an explicit save command for individual transactions. You click a Save or Record button after you have entered an individual transaction, and you cannot go on to another transaction until you have saved the first one—just as with a document-centric application. However, there is no concept of saving the entire set of data in your account as you would save a spreadsheet. The process is analogous to "saving" an individual line of a spreadsheet. What is loaded when you open the account is all of the transactions, but what is saved is each transaction as it is modified.

Database applications such as FileMaker and Access do not use documents in the traditional sense at all. There is no loading a total set of data as there is in a document-centric application. Data is loaded on an as-needed basis, and new entries are normally saved at a logical point, such as when you move to another data element. This saving is managed totally by the software.

All of this is not some esoteric theory about the ins and outs of application software design. Although both iPhone and Facebook apps can function in the document-centric architecture, they function better and more naturally without it.

If you structure your app to save data automatically at logical points, users never have to worry about saving (or losing) data. Which is why this issue is relevant to incoming mail (and phone calls): switching out of your app on a Facebook page or out of your app on an iPhone should normally not result in data loss or corruption. If something is lost, what is lost should make sense to the user and not be some random assortment of now-missing data.

Few things upset users as much as the loss of their data. You want them to trust your app, and if you destroy their data (even if you contend it is their own fault), you lose that trust. You can easily see how the viral world of the social web can cut two ways: scan discussion boards about apps to see what people don't like. You can find reviews of apps on Apple's iTunes App Store, in the Facebook Application Directory, and all over the Web. Look at the negative reviews (there are many). The most common complaints are that the app doesn't work or that it has lost a user's data.

■ ■ WHAT DATA DO YOU STORE?

Many traditional desktop applications evolved from manual and paper-based processes. The information they store is pretty much taken for granted.

When it comes to apps, the question of what data you store may not be so simple. Many apps are doing new things rather than just computerizing a manual process. Some of the data have traditional parallels: an event list on an iPhone or Facebook app is much the same as a to-do list on a personal computer (or on a paper-based pocket organizer).

Some data cannot be stored. On Facebook, for example, you can only store information provided directly by the user; you can also store that user's Facebook identifier (a number) as well as identifiers for groups, events, and the like (also meaningless numbers). You cannot store anything else, including the names or identifiers of the user's friends. You can access that information at runtime, but you cannot store it. To do so is a serious violation of the Facebook terms of service.

Many people think this is just a matter of privacy or of Facebook controlling its own data. Those are certainly valid points, but there is a much more important one: everything except the Facebook identifier can change. People can change their marital status, their address, or their name. Friends can come and go. If apps store this information, it quickly becomes out of date. Thus, the terms of service prevent apps from storing it on the combination of grounds of privacy, competition with Facebook's own data, and practicality.

The underlying principle here is the same on both iPhone and Facebook: store only the data you must. Any information that *can* be retrieved or calculated on an as-needed basis generally *should* be.

Here is a caution to remember: before storing any data, make certain you are not violating the terms of service of your environment. It may seem obvious to store the user's name so that you can provide a customized greeting, but this violates the Facebook terms of service. Get the information from Facebook when you need it. What you can store, however, is anything the user types in. That means if your app lets a user pick a name to use in a game, that name can be stored.

Apps depend on their environment to run. As pointed out several times, this relationship is mutually beneficial. You must hold up your end of the bargain to be able to continue to function as an iPhone or a Facebook app provider.

■ ■ How Do You Store Data?

Everyone knows that data—even database data—is stored in files. Files are stored on disks: these can be local disks, disks accessed over a network or the Internet, and even removable media such as CD-ROMs and flash drives.

What "everyone knows" is often wrong.

On both Facebook and the iPhone, you can store data in files as you might have done in many traditional applications. However, most of the time, you use higher-level functionality. Both the Facebook platform and iPhone provide a number of ways of storing data in addition to traditional files.

Using a Database. Many Facebook apps take advantage of a database that is accessible to the Web server that runs the app. (For many people, this is open source MySQL.) This database can be on the Web server, on another computer at that location, or even elsewhere on the Internet: it just needs to be accessible to the Web server when needed. Very large Facebook apps often use distributed MySQL databases to balance the load.

With Facebook, an issue arises when you use an external database. Facebook apps can experience rapid increases and decreases in users. In consequence, you may need to reconfigure your database quickly (and frequently). Database hosting services such as Joyent are designed specifically for Facebook apps and other uses that require this fast scalability.

On iPhone, you can access a database over the Internet for your own data storage. There, too, you can encounter the need to quickly reconfigure the database server if your app suddenly goes viral.

On an individual iPhone, SQLite, another open source database, is built in. Thus, you have basic relational database functionality available for your apps. It is basic functionality because, as its name suggests, SQLite does not support the entire SQL specification. In addition, it is designed only for a single user, which is what an iPhone has. MySQL handles multiple users.

Using the Storage APIs. The iPhone has a Core Data framework that lets you store data using procedure calls (application program interfaces, or APIs). The iPhone OS takes care of the physical storage for you. Facebook has experimented with a similar API-based data store. One of the virtues of this data storage API is that it uses Facebook's own servers and is designed to be quickly scalable.

Using APIs lets you outsource data management to the app's environment. Many developers are leery of using someone else's code that they do not control, but using the storage APIs is usually an excellent idea. It can speed up development time and reduce development cost. Remember: both iPhone and Facebook want you to build successful apps, and they help you in many ways, including giving you data storage APIs.

Using APIs also gets you out of the business of managing and maintaining a database. If you are running a database, not only do you have to worry about performance, but you have to worry about software updates. Although you can control when you install them, your flexibility is somewhat limited.

Of course, when you are using the APIs, your data storage mechanism can be modified due to changes in Facebook or iPhone. Bugs occur occasionally, but in some ways it is more comforting to know that everyone who is using the API will be hit than just having your app and its database crash.

■ ■ When Do You Store Data?

All of these threads come together as you consider when to store data. The answer to that question will depend on what data you are storing—particularly the volume. If you are using external documents (or thinking of them as you design your data storage functionality), chances are that each data storage call will store a lot of data (an entire document's worth). That can mean a significant amount of data to be sent to the database and, perhaps, a significant amount of time.

How you store the data also comes into play. When you are using a database structure, whether it is in a database you control or through an API, you can store small amounts of data easily. One of the features of a database (including an API that accesses a database) is that the stored data can be identified. Finding a name or address in a word processing document can require human intervention to scan the text, but in a database, it is a simple matter of retrieving an address field for a record with a specific ID number. Part of the power of databases comes from the fact that specific data elements can be identified and retrieved individually. Today, databases are optimized for rapid access to data elements, each of which is often small. On database-driven websites, it is easy for several dozen database calls to be executed in order to display a single page (or even part of a page).

The fact that databases can store specific data elements quickly has an impact on when you store data in your app. If you are trying to bring the baggage from traditional applications with you, leave it behind. Store data in small chunks as needed and without user intervention. This keeps your app responsive and means it will behave as users expect: it will not lose data if they switch out to answer a text message.

Note that the documentation and discussion groups for apps address performance and database issues. Read them, and experiment with different strategies for storing data. You may have to fine-tune your strategies, particularly if your app is growing rapidly or you are using a shared database.

■ ■ CACHING DATA

Once you have retrieved or created data, you have opportunities to cache that information in memory so you can retrieve it without another access to the data store. All through the history of computing, there have been trade-offs between the use of memory and the use of data storage. When memory is plentiful, you do not need to worry about storing data, except when other people need shared access to it and when you need to provide for backups and safe storage for a user who leaves your app. When memory is scarce, you rely on more data store accesses because you cannot keep all of your data in memory at one time.

Memory is now cheaper than ever, and larger amounts of it are available to apps than ever before. Use memory and cache data when needed.

However, many applications accidentally use more memory than they need. Most commonly, this happens when the application allocates some storage for a needed item—perhaps a product number—at a certain point during processing. When the transaction is stored, that product number is stored in a database or document.

In well-behaved programs, that memory location is then released. The next time that section of code executes, another product number is stored in memory. If the previous memory location has not been released, another one is created; for each transaction, memory will be allocated for a product number, and it will never be released.

This is called a *memory leak*, and it should be avoided at all costs. It can cause apps to crash for no apparent reason, and then you are in the position of disappointing your users.

■■■Working with Photos and Video

Photos and video are key components of Facebook and iPhone. Depending on the survey, Facebook is often credited with being the world's largest photo-sharing service. With iPhone, users always have a camera at hand, and of course this applies emphatically to iPhone users who are using the Facebook iPhone application.

Integrating photos and video with your app can increase its power and usability, but make certain you are not just duplicating functionality that already exists. (Duplicating Facebook or iPhone functionality in your own app is almost always a mistake: build on it and leverage it instead.)

Photos and video opportunities are slightly different on iPhone than on Facebook, but the principle is the same: integrate and use them to make your app more effective. On iPhone, photos and videos can be used to record something here and now with additional information added automatically (GPS data, for example) or by the user. There is almost no value added if you capture a photo and let the user write a descriptive caption; he or she can do that with the iPhone camera. But if your app is focused on a specific area, the photo can be used in more sophisticated ways.

One way to use photos and videos effectively is to think of your app as a front end to a database. Both Facebook and iPhone can be centralized repositories of a users' photos, and from there, users can insert photos through your app into a database. This can be part of the sharing Facebook and iPhone support, but it can also be the heart of a business site where people post photos of their accomplishments related to that company's products or services. Thinking of the app as a front end to a database means that instead of someone uploading a photo and typing in a description, you have prepared fields for the database which they can fill in. This makes for fast and consistent data entry as well as simple search and retrieval of data.

■■■ Sharing with Friends and Contacts

This is the heart of iPhone and Facebook: it is fair to think of them as a quantum leap forward in socialization and communication. If you doubt this, consider for a moment what you would be able to do with either iPhone or Facebook if you had no friends or contacts. Solitaire and crossword puzzles would be almost your only options.

When it comes to apps, both environments provide great opportunities for games like solitaire and crossword puzzles. Many people have made a lot of money this way. You can also write apps that do not involve other people: think of the many self-improvement apps that let you keep track of your exercise regimen, thoughts, calories, and garden plants.

But in general, these one-person apps do not bring a lot to the Facebook and iPhone environments. True, it can be convenient to have them at your fingertips when you are already using Facebook or iPhone, but you can have the same basic interaction just by running a stand-alone app. It is when an app enters the world of sharing that it has the potential to grow massively. Even one-person games can demonstrate this if the game uses various tools and techniques to let friends or contacts know about it and, perhaps, to challenge them to compete for a higher or lower score.

As you start to think of an app, here is a productive experiment. Seriously and realistically try to evaluate how the app will use friends and contacts, or the social graph. Look at it from both sides: not just what people can do with your app but how they can accomplish similar goals *without* using your app. If people can achieve the same results using a computer program, then maybe that is what you should be developing. If they can do it with pencil and paper, maybe you should be in the stationery business. What you normally look for is an idea that cannot be implemented at all without iPhone or Facebook. As a second choice, an idea that cannot be implemented as easily or as well without iPhone or Facebook will do, but the holy grail is something that simply must be an iPhone or Facebook app. Using the social graph in an innovative way is one of the best ways to do this.

5

Design Rules for Your App

FACEBOOK AND IPHONE APPS have unique "looks," which come from the Facebook frame and from the iPhone controls when they are visible. This chapter explores the why and what of design issues for apps so that you can make yours essential to users (and profitable for yourself).

■ ■ ■ The Advantages of Fitting In

For a long time, people have worked hard to create distinctive looks for their applications and websites. For the most part, apps eschew these efforts. There are several very good reasons for this.

■ ■ DESIGN FAMILIARITY TO PROMOTE USABILITY

When your app adopts the Facebook or iPhone user design aesthetics, it loses some of its uniqueness, and that's good. People who know how to use iPhone or Facebook can feel that they know how to use your app before they have even tried it out. All the mechanics of interaction such as scrolling, making choices from a list, setting dates or times, and paging through results data are familiar. Some of the best apps have absolutely no new design elements, but they do have one fantastic idea that extends and builds on iPhone or Facebook.

Adopting the Facebook or iPhone user interface can make it easier for people to try and use your app, but it can pose a dilemma for you. Today, many companies have a presence on multiple platforms. They have a website; they may also have an iPhone app and be developing a Facebook app. It is almost impossible for all of those presentations of the same data to share a common interface.

Nevertheless, "almost impossible" leaves room for imagination and creativity. In many ways, this is a classic issue for designers: how to achieve a unified look across different media types. For years, designers have devised strategies so that a brand translates equally well to T-shirts, highway billboards, TV commercials, and black-and-white newspaper ads.

It has taken you a lot of time and effort to create the information and functionality of your website, and you should seriously consider how to move all that work to an app (*repurposing* is the term that is most often used). One of the things to avoid is hiding information—letting people get to a certain familiar point on the website or in an app only to discover that the next logical step is available only on the other platform. Every piece of information and functionality needs to be evaluated in terms of what purpose it serves for the user and on what platform (or platforms) it is presented.

If you are one of the many people repurposing content for both Facebook and iPhone apps, you can start by looking at how others have addressed the issue of very different sizes and shapes. One common solution is scarcely new: rely on your logo. Do not, however, just copy and paste the logo from your website to an app. Different screen sizes and resolutions mean one logo does not fit all. Large companies generally have a range of logos that all look about the same but are suited for different uses.

If you have a small company, you can try to do the same thing, but you can also use a somewhat faster and usually effective method. Concentrate on what is important in your logo and reuse that. Oftentimes the color scheme of apps evokes the color scheme of the logo. The color red and cursive script pretty much defines the Coca-Cola logo the world over. You can bring many changes to the specific implementation; it is amazing how those two design elements still convey the message. Other recognizable brand images such as the Apple logo and the LaCoste alligator are remarkably stable in a variety of environments even without all of their details.

If there is any possibility that you will want to integrate your app with other apps and media, this is an issue to address early on. If you are developing or changing your logo, consider these issues during the design process even if you have no specific plans for an app yet. You can achieve a unified look by abstracting the key elements and using them appropriately. A logo or consistent design strategy—even if it differs in the details—will help reassure users that they are in the right place.

■ ■ DESIGN TRENDS OF THE TECHNOLOGY WORLD

Software has its own design trends. In the case of user interfaces, these are often driven by technology. If you compare interfaces today with those from ten or twenty years ago, you will be struck by the decidedly different looks.

User interfaces today rely on the high resolution of modern displays. This means more colors, more visual effects, and smaller objects in the interface. Twenty years ago, interfaces boasted larger objects and a much more limited palette of colors.

You can look at a user interface and date it pretty accurately. This is yet another reason to use the interface elements for iPhone and Facebook: as they change, your interface will be updated automatically. Both environments allow you to create interface elements by calls to API routines, which guarantees that when the interface changes (and it will), your app will automatically fit in. If you insert customized interface elements, you will have to do the updating.

▪▪▪Let the Environment Keep Your App Up-to-Date

On the Web, in stand-alone applications, and in the world of apps, a new interface has emerged. It reflects two major trends:

- Partial page loads
- Interfaces that are simpler to create and use

These may sound esoteric, but they have important consequences for users and app designers.

▪▪ PARTIAL PAGE LOADS

Until 1999, the smallest unit on the Web was a page. If you wanted to load any data, it was always a page (although perhaps a small one). Today, it is possible to load just a part of a page—the part you care about.

This technology is what makes many Web pages much perkier. Instead of clicking Next Page to load another page with all of its graphics and code, you can click a tab or another button so that only the changeable section of the page is loaded. Some studies have shown a 50 percent reduction in bandwidth requirements for pages coded in this way. And without being a technologist, you have probably noticed how much more responsive some websites are today.

Both iPhone and Facebook support partial page loads; in fact, Facebook's design relies on them. Without knowing what is going on behind the scenes, users have come to expect perky response times from Facebook and iPhone as well as their apps. As you design an app, you can leverage this responsiveness. One common way is to use a tabbed interface where only a specific tab's data is loaded. Whether your app is designed for iPhone or Facebook, chances are you will use (or at least seriously consider) tabs in the interface.

As you design your app, make sure you go with today's way of designing: ditch the metaphors that are a hallmark of the first generation of graphical user interfaces (GUIs). The GUI on computers is widely considered to have begun at the Xerox Palo Alto Research Center (Xerox PARC) and was popularized on the Apple Macintosh in 1984. The GUI relied on a screen display rather than text, and it integrated a mouse along with a heavy dose of metaphor.

In those days (a quarter of a century ago), the screen not only helped users navigate through their software and data, but it also gave them ideas of what they could do. As evolved by Apple in the Macintosh, the screen came to represent a *desktop*. Information was presented in *documents* that could be grouped into *folders*. The mouse could be used to move things around so that a document could be moved from one folder to another or even deleted by moving it to the trash. All of these objects—documents, folders, and a trash can—were represented by images that, in very low resolutions, looked something like their real-world counterparts.

Over time, graphics and speed have improved, so the interface is more attractive than the pixelated black-and-white version from 1984. In the 1980s, people did not know what personal computers were and what they could do with them. The metaphor-laden GUI helped with both issues. However, two problems gradually started to arise as new applications for personal computers started to be developed:

- **The office/desktop metaphor no longer fit with new applications.** If you are sitting under a tree using your iPhone to find a nearby bakery, what does that have to do with documents, folders, and a trash can? At an even more basic level, the rise of databases and the decline of document-centric applications called this interface into question. And to make things even more complicated, the metaphors for real-world objects ignored one of the key features of the digital world: digital objects can be in many places at the same time. A piece of paper is in only one place at a time, and its metaphorical representation in a GUI often observes that constraint. Getting the most out of the digital world often means handling information that is simultaneously in several places and several states (such as appearing both as an image and as editable text) although the underlying data is only one object.

- **As new applications were developed, new metaphors and icons were needed.** Developers and designers spent hours devising metaphors and icons for conceptual objects such as appointments, recipes, or parole reports. Appointments could easily be represented by a calendar; recipes by a document with food superimposed in a corner; and a parole report by a document with some other image superimposed in a corner (what that other image would be is not clear). Then to further complicate things, you needed icons for actions. What would the icon be for approving a parole report— superimposing a check mark over the document with the parole report icon in the corner? Something had to be done.

Designers and developers, along with users, have come to a consensus that people now know how to use computers. Legacy (older) software can still rely heavily on icons and metaphors, but new software—including apps—tend to use a different interface style. The fact that a great deal of legacy software still exists can sometimes prevent people from appreciating the extent to which new software such as apps needs to play by new rules, but if you want your app to be successful, you should play by the new design rules.

New Design Rules

Here are some of the specific implications of the new design environment. It is graphical, but it is not the icon-and-metaphor type of interface from the past. The design of the application's interface and functionality must evolve as a whole. In the past, it was possible to construct the functionality on its own (the functionality of a tax preparation application, for example, is pretty much defined by the tax form) and then to develop an interface to let people use that functionality.

Today, because of changes in interfaces, the interface and functionality should be designed together. Because of the limits of iPhone (mostly screen size) and Facebook (working inside the Facebook frame), you need to make certain you can expose the functionality to your users with a good interface element. At times, you may start from the interface, and when it is done, you have determined what the app does and how.

■ ■ FORGET THE MENU BARS

The first experience users have with an app is pivotal: you have a few seconds to get them started and excited. What the app does and how they can use it should be immediately apparent. Today, people are used to what is called *direct manipulation*. Want to enlarge the view on an iPhone? Use a gesture with two fingers to spread it. You do not go to a menu bar and select a command. (Which would be a neat trick on the iPhone, because there is no menu bar!)

Your Facebook app runs inside the Facebook frame, which appears in a browser. The user sees a menu bar with the browser's commands in it, and the Facebook frame includes buttons and other controls that let the user navigate through Facebook. Controls for your app are, in a sense, the third level of control.

Because there is no menu bar on iPhone and because the menu bar in Facebook apps belongs to the browser, you must implement commands in another way. Look at Facebook and notice the tiny icons such as the thumbs-up Like button, the calendar to schedule events, and even small text-based commands in a different font or color such as Comment. On iPhone, the standard E-mail, Call, and Map buttons serve similar specific purposes that, in the past, would have been implemented in the clunky menu style: select an item and then choose a menu command. Now the command is right next to the item on which it will act.

Never mind menu bars. Bring the tools to the objects that will be affected by the mouse click or tap of a finger.

▪ ▪ TAKE ADVANTAGE OF COMMON CONTROLS THAT MAKE APPS EASIER TO USE (AND DEVELOP)

Facebook and iPhone have sets of controls that apps can take advantage of; they have evolved over time and have now developed into objects that users recognize and understand without thinking of their metaphorical roots. The playing controls of forward, pause/play, reverse, go-to-first, and go-to-last can be used without hesitation or explanation, and you should never think about devising new graphical controls to replace them.

Another standard control is a hyperlink. In its text form, it is usually underlined and a distinctive color (often blue). Its meaning is clear: it takes you to another location and, most important, away from what you are looking at. This last point is not always true because in a browser, a hyperlink can open in a new window. But most of the time, it opens in the same window you are looking at. On iPhone, there is no other window, so this is a moot point. (This standard meaning of underlining to indicate a hyperlink illustrates the reverse situation: avoid using commonly recognized controls for other purposes. Underlining text on a Web page is considered bad form because people will tend to click on it thinking it is a hyperlink. The recommended practice is to use italics for emphasis; the HTML <em> element implements this behavior so that emphasis can be provided by italics or any other font style without changing the basic HTML.)

Other interface elements that can be used without worry or explanation include tabs, page forward/back arrows, and jump-to-page-number links that are self-explanatory when provided next to forward/back arrows at the top or bottom of a display. In addition, where relevant, a trash can at the bottom of a display may retain its ancient GUI function.

On iPhone, a wide variety of gestures are also built in and should be used in preference to buttons and icons. These include the calendar controls, the arrows to greater and less detail, sliders, and on-off buttons.

Perhaps the most common control is the all-purpose and unspecified action control. It may appear as a gear wheel, as it does on many personal computer applications. It is commonly understood to mean that "this control will take you to an area where you can enter data or set values for the item immediately to the left or right of the control." This is useful on personal computer applications with their wide range of commands. For apps, it is less necessary because the commands in any given context are usually quite focused and specific; the problem such an icon addresses—the ability to choose from an array of a dozen or more possible commands—is less common in these environments.

■ ■ REMEMBER APPS HAVE ONLY ONE WINDOW

Compared to a traditional application, one of the biggest differences with apps is that you do not have multiple windows open at the same time. On iPhone, this is obviously impossible. With Facebook, a user can have two browser windows open with both of them showing Facebook pages. But even if he or she is using your Facebook app in two windows, each copy of the Facebook app is separate. If your app uses storage based on user ID, the user could make changes through your app in one window that would be reflected in the other window when it is refreshed, but it would be the database—not the individual copies of the app—that forges the link. In any event, this would probably be an app with a very specific purpose. It is also possible for a Facebook or an iPhone app to open another window somewhere on the Web, but this, too, is a special case. If the intent is to return from the foreign Web page to the app, you need to open the foreign Web page in a separate browser window on Facebook.

It is possible for an app to open a transient view that can appear in front of the current window to display information or, in some cases, to facilitate data entry. These are used on Facebook in a variety of cases; on iPad, they are called popovers. They generally are clearly subordinate to the main window and are smaller than it so that they appear to be on top of it. Popovers on iPad refine this interface by having an arrow that points to the element that caused the popover to appear; they also disappear as soon as they are no longer needed—for example, when a user has tapped a choice in the popover.

This means that everything the user cares about has to be presented compactly and simply in the one window he or she is dealing with. You cannot have the luxury of multiple windows for simultaneous displays of different aspects of data (schedule, budget, comments, and so forth). It all has to fit in that one window, and if it doesn't, the information you are presenting has to be streamlined and simplified. In many ways, the visual constraints of apps force you to write better apps.

■ ■ Satisfy the User—Fast

Give your app a competitive advantage by ensuring its speed on opening. When a user wants to open your app, it should open quickly. Many times, as a consequence of the single-window display in an app, the user needs to quit your app (iPhone) or open a new browser window (Facebook) to go somewhere else. Apps are either running on the iPhone as the current app or they are not running. There is no background processing.

If you are writing an app, you should do everything you can to make it snappy. If you are hiring someone to write it for you, one of the most valuable things you can do is to invest in a stopwatch if you do not already have one (there is one on the iPhone and most computers have one, but you need one you can hold in your hand). Time everything: launch time, time to display a piece of information. Do not just look at the raw times; keep track of their variations. Whether they realize it or not, users get used to predictability in their apps. Figure out what makes the same operation faster or slower in one case or another. Then you (or your developer) can attack the problem. Remember that you do not always have to solve the problem of slow-performing processes. If you can identify operations that can take time, you can alert users to expect a delay. It takes no technical skills to do this type of performance testing.

On personal computers, users are used to an application taking some time to launch. Recent changes to the major operating systems have focused on improving launch time, and developers have started to address the issue. With traditional productivity applications such as word processors and spreadsheets, the launch time of document-centric applications often depends on the size of the document to be loaded and processed. Users know that opening a spreadsheet with ten thousand rows takes more time than launching the spreadsheet app with an empty spreadsheet and why there is a difference.

The best way to handle the loading of data and resources is to be lazy about it. *Lazy loading* (or *lazy initialization*) refers to a design pattern in which objects and data are created or loaded on an as-needed basis.

Because fast loading is critical for apps, loading data and resources on an as-needed basis is the norm. This may mean rethinking an existing application's structure as you convert it to an app. But the benefit may be speeding up launch time for the existing application as well as the app.

■ ■ USE FACEBOOK AND iPHONE RESOURCES TO IMPROVE YOUR APP

Facebook, Apple, and Apple's iPhone telephony partners all have a stake in your app running well on their platforms. Developer documentation for both iPhone and Facebook provides code samples and documentation of best practices and coding techniques. Apple provides extensive debugging and performance monitoring tools for developers. (These are the same tools that are used to debug and monitor software for Macintosh computers, including the operating system itself and many applications from Apple and third-party developers.) With Facebook, a variety of open source Web performance tools are available to help you monitor your app—which runs on a standard Web server—as well as any database(s) it might use.

The last feature of the new design environment is not new; in fact, it is one of the cardinal rules dating back to the original GUI design. Every user action should have an immediate and visible response. This applies no less to apps, and it is another example of why the user interface and functionality need to be developed together. If a user can issue a command to do something with your app, you have to think about not only the interface design that will let him or her know that this function is available and how to access it, but also how to provide feedback when the action is completed. *Feedback* does not mean popping up a dialog that says, "Your command executed successfully." It means showing the user what has been done.

This brings up two additional points.

Undo and Save. Two of the most important commands on the menu bars that are no longer available for your app are Save and Undo. For the most part, apps perform actions in response to user commands. Those actions are focused and specific; the most successful ones can be carried out on a single page (whether on iPhone or Facebook). Multipage data entry is still necessary in some cases, such as entering events and contacts on iPhone, but the sequence is controlled by the user starting from a master page that shows the subcategories (Alarm, Date & Time, and so forth for events). There is a Save button available to save all the data in these cases.

There is no undo capability. Users can delete items they have added, but they cannot go back to what the data was like just before they started to edit it. When you think through the consequences of this style of design, you can see that it places a premium on actions that are discrete and focused (or *atomic*, if you want the technical term).

Fortunately, constructing an app based on such actions means that those actions are easy to explain to users. In fact, most of them are self-evident given the domain in which the app and the user are functioning.

Failure Is an Option. Your app should not fail, but users may fail you (at least by your standards). You cannot force users to do it your way, but you can design defensively. For example, the built-in Calendar app lets you add a new appointment by tapping the plus sign (+). You can enter a location and a name and change the date or time. Yes, *change.* All new appointments begin with a start time of the next hour and a length of one hour. You cannot create an appointment without start and end times. You can change them, but you cannot delete them. All of the other fields are optional.

For the built-in Contacts app, you can tap the plus sign to create a new contact. Until you type in some data, the Save button is disabled (there is a Cancel button). It does not matter what you enter (name, address, phone number, or e-mail); no field is required as long as one (any one) is provided.

The key to your app's success is its value to users. The suggestions given in this chapter all help you deliver that value simply and quickly. Strip away anything that gets in the way of a direct and absorbing user experience. This will help make your app as good as it can be, and that helps ensure that users are satisfied, use it again and again, and recommend it to their friends. The viral spread of the social web continues.

Once you have achieved that goal for the first time, it is time to move on and position your app in the real world, which will be discussed in the following chapters. This is an iterative process that never ends: you will continue to make changes to make the app as good as it can be, evaluate the results, collect feedback from users, refine the interface (which usually means simplifying it), and so on again and again.

6

Making the Most of iPhone Location Features

YOU HAVE SEEN THE basics of Facebook and iPhone apps. Now it is time to move on to the specifics of the two platforms and the strategies for profiting from this new environment. There are great commonalities, but there are also great differences. The old saying about not putting all your eggs in one basket certainly applies to these two app platforms.

The preceding chapters have explored many of the commonalities of the two platforms. Now the focus shifts to what makes Facebook and iPhone different from one another. The two most critical differences are the iPhone's location features and Facebook's social integration tools. This chapter and the next one look at iPhone and its location features; chapters 8 and 9 address the Facebook tools.

Two sets of location features matter to iPhone app developers, and they reflect two different senses of the word *location*. The first has to do with where the iPhone is, and in most cases, it is in the user's hand or on his or her body (in a pocket, belt holster, or backpack). That location is what this chapter is about. Chapter 7 explores the other sense of location—where the iPhone user is.

For many developers and users, these features make up a new set of issues. As long as it is connected to the Internet, it does not matter where a computer is, and until now, most computers had no awareness of where they were (although a user can set the time zone and often the region). Computers have typically relied on users to set their location, and if a user moves a computer, its location information is unchanged until the user changes it.

A device that can determine its location is something new. Developers and users are still exploring this ability (called *geolocation*) with iPhone. The basics were obvious in the first few weeks of release of the GPS functionality: users can locate nearby friends and businesses as well as themselves. Now more sophisticated geolocation features are building on the underlying technology.

Part of the iPhone experience is the fact that it is close at hand for most people, and that makes it a perfect vehicle for spur-of-the-moment use of apps. This, too, is something new. Rarely does someone launch a word processing program on an impulse (save for the impulse to write a novel), but picking up an iPhone on impulse to use an app for perhaps thirty seconds is reasonable. It is your job to make the experience rewarding enough that the impulse turns into ten minutes or more and that the user picks it up again soon.

■ ■ ■ Hands-On with iPhones

iPhone is a computer, a computer that is more powerful than many computers that were on the market just a few years ago. If you think of the power and flexibility of a computer as you are developing iPhone apps, you will be on the right track. Almost anything that can be described logically can be done on iPhone. ("Described logically" is the key: you will never be able to open a can with an iPhone.)

The differences between iPhone and traditional computers have real, down-to-earth consequences for apps. Obviously, one of the first differences is that you hold the iPhone in your hand; very few people use it any other way, except for those using the speaker or an earpiece for hands-free phone calls. That is a big difference from a computer that is located on a table or desk most of the time. Even laptops are used more often on flat surfaces than on laps.

■■■ Quick—Do You Have Your iPhone?

Everyone dresses differently, but the pockets of many people contain familiar and similar objects. Backpacks and purses take up some of the burden, but many people still rely on their pockets. That can mean that keys are always in a certain pocket, a wallet is in another, and the iPhone is somewhere else. You may not even know what is wrong when you leave the house without one of these items, but you will often feel that something is wrong long before you realize that the familiar weight of keys or the bulge of a wallet is missing. An iPhone is one of those items that you may occasionally lend to someone but that you expect to have returned to you in a short time.

Many iPhone users have the phone with them at all times. In addition, particularly for those users who carry the phone on their body (not in a purse, backpack, or briefcase), its absence is often noticed immediately. You can assume that most people have their iPhone with them almost all the time, even though it may be in Airplane mode on a plane, at a performance, or in a fancy restaurant. You rarely hear people say, "I must type that into my iPhone when I get a chance." Because the iPhone is readily available, they tend to do it right away—or not at all. Apps help people do this by using techniques such as prefilled templates for e-mail and appointments to make data entry as fast as possible.

In most cases, the owner of the iPhone has possession of it. Of course, people often borrow a phone to make a specific call or join in a conversation, and just as with any other personal possessions, the borrowed object sometimes winds up in the borrower's pocket or briefcase. But for the most part, you know that the iPhone owner is also the user. With traditional personal computers, you have no similar way of assuming who the user is.

If your app is an iPhone app, it has been purchased and installed through the iTunes App Store, and that provides a form of identity verification just as purchase of a contract for use of the iPhone on a telecom network does. Although you do not have access to all of this information, it is available on request (and possibly by subpoena) to various organizations and law enforcement agencies. The former anonymity of the Internet is largely not present in the world of iPhone (and other phone) apps.

This in no way means that the actions of people who use your app are visible to the world at large or even to authorities; however, it does mean that the anonymity some people have come to expect on the Internet does not exist. For most people and most apps, this does not matter, but for apps and people who may be functioning on the fringes of society and the law, it is a serious concern.

For the vast majority of apps and people who are not skating on thin ice, this means that the iPhone environment is somewhat safer and more secure than the no-holds-barred areas of the Internet. Your app should not encourage people to evade the licensing and terms of service restrictions (which, in any event, would be illegal).

Many people believe that the absence of anonymity on iPhone (and Facebook apps) is a good thing. When you add a Phone button to an iPhone app, you should assume that the user will make a call and that caller ID may identify him or her to the recipient. (The user, of course, should also make this assumption.)

■ ■ ■ Apps Are for Spur-of-the-Moment Use

Because users do not have to go to the iPhone the way they go to a computer on a desk, it is there at the moment they think of using it. If people think of using your app, they can use it right away without a complicated start-up process—if you have done your job properly. Above all, you need to make the app's initialization speedy. If initialization relies on checking with a remote server over the Internet, you can have a significant delay. Cache this data locally in the iPhone's database so it is available and the user can get started right away. You can always connect to the Internet in the background to update the information.

What triggers someone to use your iPhone app? Some of the triggers are the same as those that inspire people to use applications on laptop and desktop computers: information in e-mail and on the Web, as well as the whole range of media people are exposed to every day.

But iPhone is not only a computer. It is also a camera, a music player, a portable Internet device, and, yes, a phone. A wide range of events bring the iPhone to mind and to the user's hand. At that point, your app has a chance to be used if it is available and if the user remembers it is there. This is one of the areas in which iPhone apps and iPhone Web apps differ.

■ ■ How People Use iPhone Apps

People can travel two routes when they get the idea to use your app:

- They can download and install it.
- If they have already downloaded and installed it, they can start to use it immediately.

That much is self-evident, but it poses an issue for you: how do you get your app on the user's iPhone before he or she thinks of using it? This is a subcategory of the more general question of how you get people to buy and/or download your app. (Even if it is free, they have to download and install it.) This is a common issue for certain marketers whose products may not be used immediately. A container of milk is likely to be used within a few days, but a jar of pickled kumquats may sit on the shelf for years.

As with many products, your marketing strategy may be developed along with the product; this applies to small-scale apps as well as the largest ones. You have to surmount the dual obstacles of purchase and installation as well as use, and your promotional and support materials have to accomplish both jobs.

There are two basic financial models for iPhone apps (additional models are explored in Part 2):

- You can generate a revenue stream from sales of the app.
- You can generate a revenue stream from sales of add-ons and enhanced functionality, often with a periodic fee (monthly and yearly subscriptions are common).

If you are basing your projections on the first model, sales are critical; in the second model, it is use that is critical.

■ ■ How People Use iPhone Web Apps

With iPhone Web apps, there is no download, no installation, and most of the time, no payment. People can start to use your app just by typing in a URL or tapping a link. The parallel to the installation process for iPhone apps is bookmarking your iPhone Web app's URL and tapping the button to add it to the Home screen or to the bookmarks list, and even this is optional. In many ways, this can make your iPhone Web app more accessible than iPhone apps (and Facebook apps).

When it comes to revenue streams, you normally do not sell the Web app. However, nothing prevents you from placing the app behind a portal and charging people for access to it, so you can theoretically charge for its use. (This technique also works well for in-house apps that access private databases using the iPhone's inimitable interface.)

All of the revenue streams available for iPhone apps are available for iPhone Web apps, and as you will see in Part 2, integrating the iPhone Web app with existing e-commerce sites can be easier than integrating an iPhone app.

■ ■ Building App Awareness

People need to be aware of your app at times when it can be useful. From the moment you first start thinking about your app, focus as much as possible on its specific benefits to users. By being specific, you make it easier for potential users to realize that your app is just what they need at some particular point. Once your app is focused in this way, you have a story to tell to potential users. Whether you are launching a major advertising campaign for your app or relying on a brief description in the app directory, your story has to be clear and specific, letting users know what they can do with your app. Even if your app's benefit is not immediately needed, the more specific your message is, the more likely people will be to download and install your app for future use.

Another way of getting people to use your app is to get a partner who urges or requires people to use your app. Such partners are varied. Teachers have used certain apps in their classes. Some apps can be used as front ends for databases, and that can make them useful (or even required) for entering data.

If you have made your app valuable, it can be offered as a gift to members of a group. In such cases, you can provide basic functionality to anyone and a members-only feature. For example, an app that helps people manage collections (such as antiques or wines) has obvious value to specific groups of people, and they may appreciate a free version from which they can upgrade to a more complete one.

■ ■ ■ Scaling the Corporate Wall—from Inside or Outside

There is a tremendous market for iPhone apps that take advantage of the iPhone's constant availability to the user. When most people think of apps, they think of those in the App Store and iPhone Web apps that are listed on Apple's site and all over the Web.

Other apps represent a market that is potentially even larger than the huge customer base for these readily available apps, and it is a market that Apple is trying to develop: the corporate world. One of the great things about the App Store is that you can see what people are doing as developers and as purchasers/users of apps. You can see what is selling on the lists of most popular apps, but most of this information from Apple comes out in press releases and can sometimes be incomplete. Nevertheless, you can find out a great deal about this public market.

Before iPhone, there was BlackBerry from Research in Motion (RIM), and it remains the smartphone of choice in much of the corporate world. Early on, RIM made design decisions that made it easy to integrate BlackBerry with corporate databases and IT systems. A challenge for iPhone and iPhone app developers is to make inroads into this corporate world. It is enormous and lucrative.

One of the reasons it is lucrative is that, in many cases, it is closed. What a corporation does with its IT infrastructure is rarely revealed to the public unless it is part of a glowing article about how the organization has solved a problem or gained a competitive advantage (or, conversely, it may be revealed as a result of litigation). In this private world, vast amounts of money have been spent on integrating these devices and apps with the corporate IT infrastructure. Senior executives who resisted doing what they considered data entry tasks on personal computers have often been wooed into doing exactly the same tasks on a phone. With large amounts of money invested in these integrated systems and deploying devices across an organization's workforce, it is hard to make a change, which is why these markets are attractive to vendors such as Apple.

In some surveys, IT managers have cited this cost of making changes as the main reason for staying with Windows PCs rather than moving to Macs, even when there is internal demand for Macs. To be perfectly fair, the one-time cost of a changeover can be enormous; even developing a multiplatform environment is daunting as managers view a doubled maintenance environment for two operating systems.

The fact that many corporations are entrenched in integrated systems that work should not discourage you. Indeed, it should excite you. Remember that the cost of entry to the world of iPhone app development is relatively low. The early adopters of technology that integrated smartphones with corporate IT infrastructures have proven that it works and is valuable. The fact that some large organizations may be spoken for in no way diminishes the value of integrating iPhone or BlackBerry with corporate IT infrastructures, so the market is wide open for the vast number of for-profit and nonprofit companies that do not have such systems. Arguably, iPhone—which is built on a different operating system than BlackBerry and has benefited from watching how BlackBerry and others have succeeded and failed in various ways—can be a better choice for a new integrated system.

Can you sell a corporation an app that handles time billing? If your expertise is in this area, you can probably develop a customized solution quickly and easily. The many time-billing apps in the App Store have varying degrees of customization, but many large corporations need much more, and that is where you come in.

You can easily build a customized expense account app that can get rid of all those receipts and monthly credit card statements. People have their iPhones available almost all the time. Instead of receipts and credit card statements, let them photograph the bill (it is usually not necessary to photograph the food at an expense account meal, but that is a corporate choice). The resolution on the camera is sufficient for this purpose, and the fact that the photo can be checked immediately—before the receipt or bill is thrown away—adds to the convenience.

■■ Advantages of Partnering with a Corporation

While you are thinking about the corporate world, think of opportunities for custom apps for a specific company. Once again, the low cost of entry to the iPhone app world, along with the fact that people usually have their iPhones available, is to your advantage.

One of the most difficult tasks for a company to manage is the registration of product sales. In some cases (the iTunes App Store, for example), the purchase of a product generates its own information about the purchaser. In other cases, such as the purchase of an ink-jet cartridge, a household appliance, or even a sweater, this information makes it back to the manufacturer only if the user fills in a registration card online or on paper.

Enter the iPhone registration app that you can develop. How easy can registration be if the app walks the user through taking a photo of the sales receipt (which has price, date, and vendor) and product code, then sending the information to a corporate database?

This is just one of many types of tasks people will carry out if made simple, and the presence of the iPhone and your app can make it simple.

■■■ Tracking iPad Location Features

The iPhone OS powers both iPhone and iPad; as noted previously, this means that, with certain exceptions such as telephony on iPhone and a larger screen size on iPad, both devices function in the same way. One distinction that is likely to emerge as iPad enters the market is that iPhones will remain close at hand for most users but iPads may not. For many people, their first view of someone using an iPad was at the 2010 Grammy Awards when Stephen Colbert reached into his tuxedo jacket pocket to bring out his iPad containing a list of nominees. Whether or not iPads become an essential part of formal attire, chances are that they will not be found on belt holsters or in pants pockets. As part of your continuing observation of how people use apps, watch how the iPad begins to fit into users' lives, and then think about how you can add value with apps.

This chapter has explored the opportunities for apps that are related to the iPhone's general availability. The next chapter explores how you can take advantage of this is another way: by using the user's location in your app.

7

Discovering App Opportunities
for iPhones

Because people often have their iPhones with them, these devices can be anywhere their owners are. Although the iPhone is a phone, a computer, a camera, a music player, and an Internet device, it is not subject to the limitations that some of those devices have. Computers are typically in offices or on desks; even laptop computers have an aura of office about them. When thinking about apps, you may have to constantly remind yourself of this so that you can identify and take advantage of the opportunities for iPhone apps that are unavailable to devices with a more limited range of locations.

Three aspects to an iPhone user's location suggest important opportunities for apps:

- iPhone is not an office- or desk-based device.
- iPhone users can be in a wide variety of locations anywhere in the world.
- Built-in GPS for the iPhone (starting in iPhone OS 2.0) lets iPhone users locate themselves and others through apps.

Note that some of the opportunities based on user location are similar to and build on those discussed in the previous chapter—opportunities based on where the iPhone is in relation to the user (pocket, backpack, purse, and so forth).

An iPhone Is Not Office Equipment

iPhone and other smartphones are new types of devices. Although they are in fact computers, and despite the fact that they have many things in common with gaming devices such as Xbox and Wii, they are in their own evolving category. When people start to think about developing apps for iPhone, they often think of programming traditional applications, but there are big differences.

Perhaps the most important is that iPhone has a very different shape from a personal computer (or even a computer terminal linked to a mainframe). It is not designed to be used on a desk or table or even resting on a lap. It is the ultimate portable device because it is not tethered by a power cord or data cables, and its communication capabilities include Bluetooth (short range), WiFi (medium range and higher volume), and a telephone network for both voice and data. iPhone users can and do move around while they use it. Think of people on the go and in a multitude of environments as you consider your app.

Who Uses iPhone?

Not only do people frequently have their iPhones with them, but they belong to a much different group from those who use traditional computers, telephones, and even cell phones. There are many people in the world who have never sat at a desk in an office, but many of these people have used a telephone. Particularly before the widespread use of cell phones, there was a clear distinction—at least in the popular perception—between heavy users of telephones (think women and teenagers) and others. Any underlying reality of phone users, however, is being changed by the use of cell phones.

Since the early days of the Web, research has been conducted to see who does and does not use the Web and other Internet tools. The trend has continually increased around the world, but there have been disparities in both access and usage. A "digital divide" between Internet haves and have-nots continues to exist, although it is narrowing. In the United States, this ever-decreasing divide has clear correlations to age, race, and economic levels.

The actual numbers and demographics of iPhone users are not public knowledge; what Apple releases is most often advertising that, quite legitimately, describes the audience of users the company would like to have. However, you can do your own market research very easily: look around. Who is using an iPhone? Who is using old-fashioned (nonsmart) cell phones? It is often easiest to bring your app to existing iPhone users.

Considering the demographics of people who use iPhones and the Internet and those who do not, you can identify distinctions in terms of age, income, and other characteristics. Look up "digital divide" on a Web browser and see what some of those characteristics are. Several trends come together here; the digitally uncomfortable grew up and lived in an era of offices and corporate structures that is much less pervasive today. The idea of conducting business in a nontraditional space or time is no longer the radical notion it used to be.

There is an enormous opportunity to develop apps for people in their middle years and older. Because of advertising and marketing (as well as a kernel of truth), many people think of iPhone and even computers as part of the younger generation. Nevertheless, the rapidly growing ranks of aging baby boomers represent a new and interesting market for iPhone apps. If these people are your target audience, do not treat them like teenyboppers. Chances are some of them use iPhones to be trendy or even to appear younger, but iPhone apps that are useful and respectful can find a welcoming audience in this underserved demographic.

Because iPhone is such a personal device, make certain your app reflects your user's point of view. It is not unreasonable to think that the collection of apps a person installs on an iPhone defines that person's interests as well as how he or she thinks. Given the thousands of apps available, it is not hard to find apps that reinforce a person's beliefs and interests, as well as apps that address the study and research needs of people who are interested not so much in reinforcement as in challenges. You probably haven't thought of iPhone and apps as personality tests, but they can be. Your app can be wildly successful if it does something people want and—very important—in the way people want it done.

■ ■ TYING THE INFORMATION AND OPERATIONS TOGETHER

Despite the trend toward moving away from offices and many existing corporate structures, the fact remains that both will be with us for a long time to come. One of the ways iPhone can be integrated into the office world is with tools that allow it to share data with office and personal computer systems and databases. Using Exchange Active Sync, as well as the synchronization tools on the Mac (with bridges to Web services), iPhone fits right in. Data is not held hostage behind a wall. If you are working on an app that will fit into an office environment, make certain you are using the synchronization tools to improve your data flow into that corporate IT infrastructure. iPhone apps can be personalized tools that let people use these corporate systems in the ways they want.

■ ■ ■ iPhone and Its Users Can Be Anywhere

Because iPhones can be anywhere, a big opportunity exists for apps that are geared to specific types of spaces. These fall into two broad categories: informational apps and transactional apps. We will examine both in this section, but the starting point is simple. When someone is using an iPhone, that person is somewhere in the world (and that location can be found on his or her iPhone). Wherever the person is, he or she can see (and hear, taste, touch, and smell) the surroundings—the furniture, the landscape, the people. iPhone apps can provide that person with information about or help him or her interact with those surroundings.

■ ■ INFORMATIONAL APPS

These apps provide information about a specific place or type of place. They include tourism apps and nature guides. If a user is crouched behind a bush with binoculars on a bird-watching adventure, iPhone is ideally suited to helping that user identify the wildlife he or she sees. Indeed, field guides that include birdsong are available through the App Store for the classic Audubon and Peterson guides.

The ability to integrate iPhone photos with information resources has not yet been fully explored, and it, too, represents a large opportunity. In general, if you are a traditional content provider, particularly in a specific context such as travel or nature, you can probably repurpose your resources for iPhone.

■ ■ TRANSACTIONAL APPS

Transactional apps are those that help people do something. This is also a wide-open area for people who get out and about with their iPhones. For many years, small handheld devices have been used for tasks like reading meters. When digital cameras were introduced, many people used them to document a wide variety of events. iPhone provides an opportunity to integrate photos with notations for everything from building inspections to in-store display documentation.

If you are involved with any operation that involves on-site documentation, look at iPhone. Granted, an adept iPhone user who understands the specific needs of a documentation task can assemble text and photos without the use of an app. But opportunity for development does not arise from supporting the sophisticated user; it arises from supporting the domain specialist or even the unskilled worker who knows the task and the vocabulary of the task at hand. Your eyes may glaze over at the possibility of selling hundreds of thousands of copies of your app to people doing on-site investigations, but the developmental work can be split into domain-specific apps that together account for a large market.

As you browse the App Store, click on the links to other apps from an app's developer: you will see clusters of domain-specific apps from people who have learned the domain and iPhone technology. Because iPhone has built-in communication tools, you can integrate existing databases and IT systems easily with both iPhone apps and iPhone Web apps, eliminating a separate synchronization process.

An iPhone app with a modest price tag (perhaps even up to $20) can be incredibly valuable to someone who uses it only once a month. In addition, if the app provides integration with an IT infrastructure at either a home company or a service provider, you have found a significant revenue stream that may include a monthly subscription fee. In such cases, you might give away the app—at least in a limited version—and then charge the subscription fee.

■ ■ Users Can Be Nowhere

Not all apps interact with the user's current location. In fact, many apps are designed to metaphorically take the user out of his or her current location. These are called *immersive apps*, and the most common example is games. The user may become lost in a fantasy world, an adventure, or even a work-related scheduling issue. In some cases, these apps are both immersive and transactional, as is the case with a scheduling app; the point is that your attention is no longer focused on where you are.

■ ■ ■ You Can Take Advantage of GPS Functions

The GPS functionality in iPhone OS 2.0 opens up an enormous range of opportunities for apps. This is an area that has already been explored by many developers, but there is room for many more.

GPS is used primarily in two different ways on iPhone apps:

- **Here and there.** iPhone can always show you where you are ("here"). It can also map any other location ("there") and show you what is nearby. You can even use it to map a route between the two spots (or between two other spots).
- **Who's there?** A third-party app can let you find nearby people with iPhones and the same app you're using.

This section gives you some examples of the opportunities to be found in both of these categories. Despite the fact that GPS applications make up one of the most popular and rapidly growing categories of iPhone apps, there is room for many more. Because iPhone combines location with the iPhone owner, any app that leverages those two elements has a built-in audience—if only you can reach it.

■ ■ HERE AND THERE

The first way GPS is used for app development is through the primary function of here and there. Apps can integrate current location information with stored data such as maps and tagged images to show the user what is nearby. The nearby items can be static in the iPhone's data store, or they can be retrieved from the Web so they can include time-specific information (such as the hours a farmer's market is open today and for the next three days).

Apps let you limit your search to specific types of locations. This was one of the first uses of GPS in handheld devices, and it remains popular, in part, because it is so easy to explain and use. Also, once a developer understands it, building the app is easy. One of the first commercials for iPhone showed a restaurant locator app that included a button that let the user call for reservations.

If you are sitting on a database of information about geographical entities (everything from national parks to restaurants), displaying that information on iPhones is a natural and obvious step. Regional tourist and chamber of commerce groups, as well as local media, can find themselves with static descriptive information about local sites and calendar information about what is happening. Such a group, too, is an obvious candidate for developing an iPhone app (and particularly an iPhone Web app).

The opportunities for these apps go far beyond tourist attractions and places to eat. There are apps to find Catholic masses and Jewish minyans based on where the user is. At aastepsaway.com, the user can find nearby meetings of twelve-step fellowships such as Alcoholics Anonymous and Narcotics Anonymous.

Another way apps use the GPS function is to locate people nearby. Users cannot locate other iPhone users directly from their iPhones; they need an app to connect them. Users of these apps must allow the app to access their location, so it can know where they are. Typically, these apps require users to register, but not all of them do.

An app that links people is normally based on some common interest—dating was one of the first examples, but you can build an app that lets you connect people with any shared interest. You can create a far-flung group of people with common interests using any number of Internet tools (not least of which is Facebook), but iPhone makes it possible to connect with people who are actually nearby.

Everyone understands how this can work with dating (at least on the conceptual level), but there are many other reasons for finding people with shared interests. One of the keys is authenticating people and knowing who you are meeting. Organizations with large memberships (such as the Rotary Club or Kiwanis) provide this sort of authentication, as do political groups and membership groups of all kinds.

As already noted, this can all be done by Facebook and other tools, but the difference is that iPhone can report users' current location. If a user is traveling, his or her Facebook hometown may still say Seattle, though the user's current location might be Memphis.

For someone who makes house and office visits, an app like this can be extremely helpful by letting clients see where the service provider is at any time. (Of course, that degree of reachability may not be what your users are looking for!)

To sum up, iPhone and its apps (including yours) are usually wherever the iPhone user is. You need to throw away years of thinking about computers as things that are tied to desks and envision people using your apps wherever they are—be that crouched in a swamp, on top of a roof, or even in an office. Think of the range of stimuli that bombard people when they are out in the world and about how your app can help them make sense of and manage those stimuli.

8

Joining with Facebook to Provide User Value to Your App

IF YOU WANT TO develop, promote, market, or use Facebook apps to improve your bottom line, the opportunities in this environment revolve around Facebook's *integration points*. These are sections of the Facebook application program interface (API) that developers can access to interact with Facebook's inner workings.

Most of the time, your app runs within the Facebook frame, taking up the entire center of the user's Facebook window. But when you use the integration points, you and Facebook jointly take over. For example, when a user is presented with a window offering to send invitations to friends, Facebook puts up that window and displays the photos and names of the friends. Facebook has written the code that lets the user select any of the friends shown in the window to receive an invitation, and the user does not have to learn anything special for your app.

But your app participates as well. When it uses an integration point such as invitations, your app asks Facebook to put up that window so the user can make choices. Your app provides basic text that can be sent with the invitations (users can add their own comments to that text). As you will see in this chapter, using these integration points is simple for both you and your users. However, it does mean that for each integration point, you need to provide a little bit of linkage to Facebook so that it can do what it needs to provide a seamless interface for users to send out invitations, e-mail messages, and the like.

It is possible to build an app that simply runs on Facebook without using any integration points, but the integrated Facebook experience is available to users only when an app uses them. Furthermore, if you are going to want Facebook to verify your app, which gives it higher visibility and a clear flag to potential users that it is officially verified by Facebook, you should know that use of the integration points is a key part of Facebook's evaluation.

This chapter and the following one explore integration points and show you how to integrate your app tightly with Facebook and why this integration matters so much. (In a nutshell, using integration points benefits both you and the user because they provide key parts of the social experience for a minimal amount of effort both on your part and on the part of the user.) This chapter focuses on built-in integration points; Chapter 9 examines the integration points your app can add.

■ ■ ■ Understanding Integration Points

Facebook has grown and evolved quickly. Its first versions did not support third-party apps, but its architecture clearly allowed for the implementation of apps in a structured and orderly way, because some of the major Facebook features themselves were implemented in that way.

As support for third-party apps has developed, Facebook has allowed developers more and more access to parts of the Facebook environment so they can use the features that have already been created and that people know how to use. It seems that every few months, Facebook rolls out yet more functionality to enhance the user experience and make app development that much easier. Just as with iPhone, letting developers leverage existing functionality and letting users apply their experience with the platform makes for an easy-to-develop and easy-to-use app environment.

Not every integration point is relevant for every app, and not every Facebook user uses them all. However, in general, the more tightly your app is integrated with Facebook, the more successful it will be and the more profitable it can be for you. These profits come from two sources:

- You provide more and more social functionality to your users while writing less and less code yourself (because you use Facebook's code).
- The social web functionality has to do with invitations and communications so that you spread your message far and wide throughout the social network.

Technically, integration points provide access to Facebook features that apps and users can then take advantage of. In practical terms, integration points are almost all about communication of one sort or another. You can divide these communications into three general categories, each of which has potentially different impacts on your app's success and functionality:

- *User-to-user communications* are communications from a specific user to one or more other specific users. In nearly all cases, the recipients are friends of the person who initiates the communication. There are almost no restrictions on user-to-user communications, and they are the most potent type of social communication precisely because they are so specific—going from one friend to another. It seems probable that these communications are most likely to provoke a response or a subsequent action (such as trying an app).
- *User-to-world communications* are centered around items on a user's profile; the most common of these communications is a user's status, which is set in the "What's on your mind?" section of the profile. Subject to security settings, your profile is visible to your

friends and general Facebook users. Parts of it may also be visible to search engines so that people without Facebook accounts can still see that information. Some user-to-world communications can be generated by apps on behalf of a user, but there are important restrictions on how an app can do this. The primary rule is that a communication must truly be sent by a specific user. This means that even if an app is involved in formulating a message, the user must still actively send it, even if the action is only a simple click of the mouse. These communications may be less potent than user-to-user communications, but that in no way means you should turn up your nose at them. Because they can be so broad based, they just might pique someone's interest. Many people find that their real-world social networks contract somewhat over time, which is why tools such as Facebook are so welcome: they help people reconnect with friends and keep their social networks vibrant. A general posting from a friend who uses your app may rekindle a friendship (and earn your app everlasting gratitude).

- *App-to-user communications* are just that: messages to a user from an app. To preserve and protect the Facebook environment, there are restrictions on this type of communication also. This category includes the news feed in which users' actions are aggregated into a set of stories describing what they have done on Facebook and with its apps. Because these stories are generated automatically, they have the highest chance of serendipitous results that can amuse and interest friends. Who would have thought that so-and-so did such-and-such, and what else has been going on these last eight years? Your app can be at the center of such interactions and questions.

As well as looking at these three types of communication, you can explore the integration points from Facebook's point of view. This is how your app actually manages the communications. There are three main areas:

- **Profile.** This is the area on Facebook where users post personal information so others can see it (subject to security constraints). Apps can post on users' profiles. Make the most of this by offering to do so and to post interesting information. The fact that someone uses your app may be terribly exciting to you, but as a profile tidbit of news, it's a bit lame. The more specific your app-generated posting can be, the more people are likely to react to it.

- **Feed.** This is the collection of news items a user can see. There are two versions: the feed of the user's actions and selections from the feeds of the user's friends. Apps usually generate these stories, and Facebook selects and aggregates them into the feeds.

- **Requests and sharing.** Users invite or ask friends to use an app or join in some other endeavor. The familiar Share button also falls into this area. Apps often facilitate all of these requests and invitations.

There is yet another way of looking at integration points. Some, such as the profile and feed, are built into Facebook. Your app can access them chiefly by providing updates that reflect users' actions. In addition, Facebook lets you extend its structure with app-specific elements, such as an entire box on the profile for your app to update or a tab for your app's use right next to the built-in Wall and Info tabs. Inside your app, you can add buttons and links to let users invite their friends to use your app.

The sections that follow describe the profile, feed, and request and sharing areas. You will see the types of communication available in each, as well as get an idea of how you can use them in apps that you build to implement your objectives.

Note that all of these features have evolved over time. If you have used Facebook in the past but do not currently use it, you may not know all of them. It is also quite possible that the features have evolved in subtle ways without your noticing, so be sure to do your homework and keep current with new Facebook technologies.

Also be aware that users can exercise a great deal of control over who and what apps can access their data. In every description of Facebook data in this book, remember to mentally insert the phrase "access is subject to user privacy settings."

■ ■ ■ Help Users Enhance Their Profile with Your App

Facebook has long described the profile as "the online representation of the user's real-world identity." This goes to the very core of the site (not to mention the users). Although many users do not do so, they can control the visibility of most of the profile information so that the representation is shown differently to friends than to others. One of Facebook's main selling points is that its users are real people who are identifiable in the real world. Posting incorrect information is a violation of the terms of service in most cases, and you can find yourself and your app out in the cold. Facebook reserves the right to cancel violators' accounts. Nevertheless, some profile information can be wrong or misleading. Perhaps the most common inaccuracy is where a person lives. Whereas with the iPhone's GPS capabilities you can rely on a person's location being correct, the location information on Facebook may be outdated. The same applies to information such as employer and relationship status (although that last category is sometimes out-of-date for reasons other than simple carelessness).

Apps are not allowed to extract data from Facebook for this and other reasons. If an app has access to specific data, it gets it at run-time. That way, even if old information appears on a user's profile, it is not cached by the app. This minimizes the potential for proliferating misleading information, but you must always remember that the possibility does exist.

Figure 8.1 shows a user's profile as the user sees it after he or she has logged into Facebook and clicked the Profile link at the very top of the window. If you are allowed to see a friend's profile page, you see a slightly different version.

In the first version of Facebook that allowed third-party apps, the profile page was different. Ads appeared at the bottom of the left column, and information in the main section of the page was laid out differently, with controls to show and hide sections such as the mini-feed, the wall, and the user information and application sections.

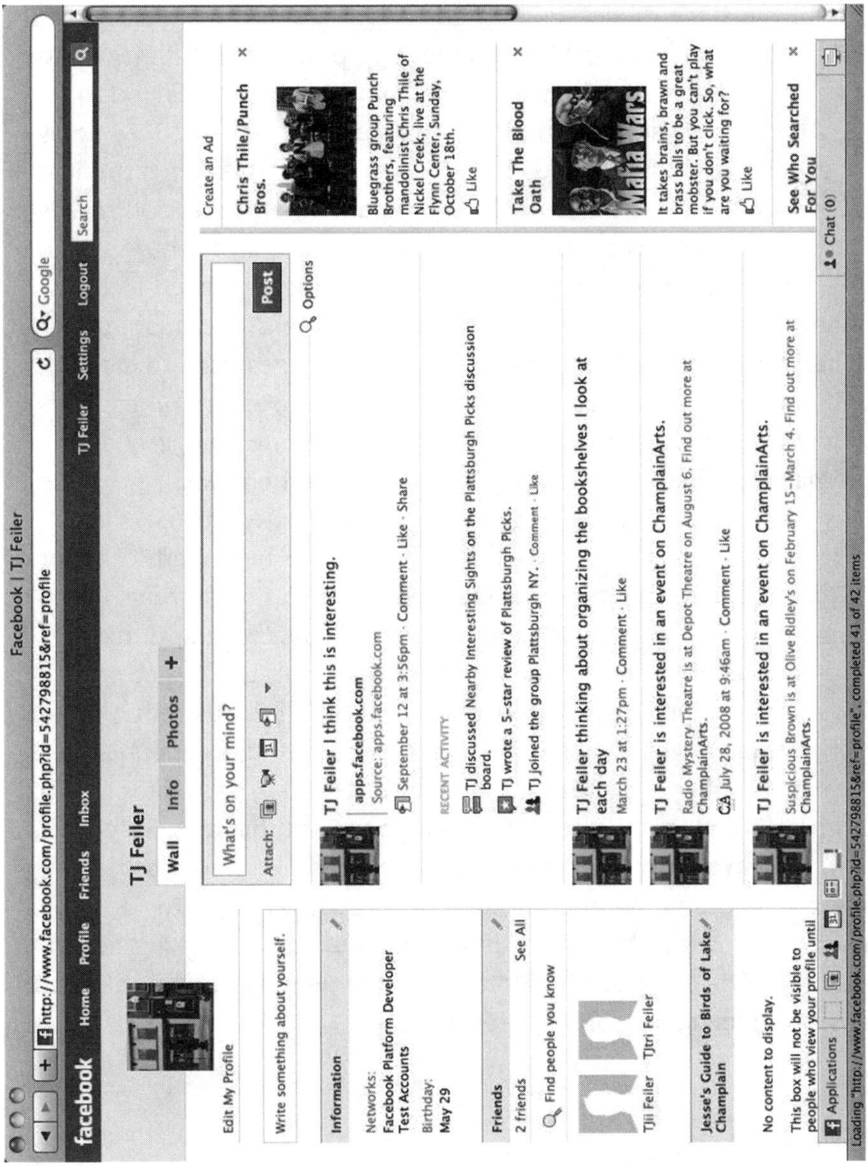
Facebook | TJ Feiler
http://www.facebook.com/profile.php?id=542798815&ref=profile
Google
facebook
Home
Profile
Friends
Inbox
TJ Feiler
Settings
Logout
Search
TJ Feiler
Wall
Info
Photos
Edit My Profile
Write something about yourself.
What's on your mind?
Attach:
Post
Options
Information
Networks:
Facebook Platform Developer Test Accounts
Birthday:
May 29
Friends
2 friends
See All
Find people you know
Tjii Feiler
Tjtri Feiler
Jesse's Guide to Birds of Lake Champlain
No content to display.
This box will not be visible to people who view your profile until
TJ Feiler I think this is interesting.
apps.facebook.com
Source: apps.facebook.com
September 12 at 3:56pm · Comment · Like · Share
RECENT ACTIVITY
TJ discussed Nearby Interesting Sights on the Plattsburgh Picks discussion board.
TJ wrote a 5-star review of Plattsburgh Picks.
TJ joined the group Plattsburgh NY. · Comment · Like
TJ Feiler thinking about organizing the bookshelves I look at each day
March 23 at 1:27pm · Comment · Like
TJ Feiler is interested in an event on ChamplainArts.
Radio Mystery Theatre is at Depot Theatre on August 6. Find out more at ChamplainArts.
July 28, 2008 at 9:46am · Comment · Like
TJ Feiler is interested in an event on ChamplainArts.
Suspicious Brown is at Olive Ridley's on February 15–March 4. Find out more at ChamplainArts.
Create an Ad
Chris Thile/Punch Bros.
Bluegrass group Punch Brothers, featuring mandolinist Chris Thile of Nickel Creek, live at the Flynn Center, Sunday, October 18th.
Like
Take The Blood Oath
Mafia Wars
It takes brains, brawn and brass balls to be a great mobster. But you can't play if you don't click. So, what are you waiting for?
Like
See Who Searched For You
Applications
Chat (0)
Loading "http://www.facebook.com/profile.php?id=542798815&ref=profile", completed 41 of 42 items

Figure 8.1

Ads have since moved to the right column, and the lower section of the left column has been freed up for app profile boxes if they exist. Profile boxes are stacked beneath the Information and Friends sections. Although these are significant changes, the fact that they apply to all pages and all apps tends to minimize their impact. It is not just your app's pages that suddenly look different when Facebook makes changes. Over time, changes to the Facebook page layout have generated a lot of discussion, including several groups dedicated to bringing back the old styles. Typically, within a few weeks, the hubbub dies down and few people even remember that the pages used to look different. Before long, the new look becomes the old one, and when Facebook makes some additional changes, a new group will likely spring up demanding that Facebook put things back to the new old way. It is always the functionality that Facebook offers that interests users most.

Once a user has added your app, it can also be added to his or her profile page from either the app or the application settings in the Settings menu. A great many things are available to an app that has been added by a user, so it is a high priority not just to get people to look at your app's listing in the Application Directory, but also to add it to their pages. Once it has been added, it can generate data for users' profile pages without any effort on their part, although it should really reflect their recent interactions with that app. All of this gets your app into the social networking graph of a user's friends, giving it a whole range of communication possibilities, provided the user has added it to his or her page and your app has taken advantage of this. Once the code is written and the user has added the app, the communication is totally automated.

■ ■ ■ Wall Tab

The Wall tab, shown in the center of Figure 8.1, includes a news feed as well as an interface element called the Publisher, both of which are described in this section. The common thread is more social interaction and, if you choose to use it, automated communication to users' walls. It is the same story over again: once a user has added your app and once you have written the code, updates happen automatically and can be propagated through Facebook's social graph to a user's friends.

■ ■ FEED

The feed contains news stories, as shown in Figure 8.1. This is one of Facebook's most important features for users. The way in which a user's feed is constructed is a carefully guarded Facebook secret, but the site clearly spells out certain parts of the process. Facebook attempts to construct a feed that contains the most interesting stories possible for users. To do so, it keeps track of stories that users have marked with the Like button, so it can give a higher ranking to similar stories. It attempts to aggregate stories generated by apps, so that if an app creates a story saying User A has done something on that app and another story saying User B has done something on that app, an aggregated story saying Users A and B have done something on that app can be generated automatically. The more friends who have done the same thing, the more interesting the story can become.

Stories often contain links to the app, to friends, or to other locations, even if they are off Facebook. These links can help you accomplish your app's goals, whether you want to increase traffic to your website, get people to use your app, or encourage them to go off-site to view and possibly buy products. In each of these cases, the first step is to get your app added to user pages and get permission to post to the news feed. Then your app has to play by the rules so it will be selected for the feed. Facebook limits the length of each user's feed so it is always manageable. One user's interaction with your app (particularly a user with a lot of friends) can automatically propagate knowledge of your app as well as of whatever goal you are trying to accomplish through the social graph. If you understand this and implement your app according to the rules, you will start to have users generating the sort of buzz that publicists dream about.

The Publisher allows users to enter brief text items and incorporate photos, events, and links with the buttons beneath the text entry box. The downward-pointing arrow brings up a menu with a list of installed apps that support the Publisher. When the user selects an app, some app-specific options become available; with the click of a radio button or some other simple interface element, the user can interact with the app and automatically generate Publisher data that will appear on the feed as well as (possibly) those of his or her friends.

■■■ Info Tab

The Info tab, shown in Figure 8.2, categorizes information into sections such as Basic Information, Contact Information, Groups, Pages (those the user has added), and the apps that have been added (and information from them).

The user can add your app to the Info tab. Figure 8.3 shows an app that contains a button for this purpose (Add to Info) as well as a link to invite other people to use the app. Invitations are another way of profiting from the social graph, as discussed earlier.

The Add to Info button is an interesting example of how Facebook manages integration. The app developer places this button in an appropriate place on the app's interface, and Facebook takes it from there. It draws this button if the app has not been added to the user's Info tab. If the app has been added, the button is not drawn. Once the user has clicked the Add to Info button, he or she can confirm the choice, as shown in Figure 8.4. Thereafter, whenever your app generates information to be added to the Info tab, the user has a chance to keep or remove it, as illustrated in Figure 8.5.

In all of these cases, the keys are the same: get the user to add your app to his or her account, and then allow him or her to take further action, such as adding the app to the Info tab. Be prepared to use all

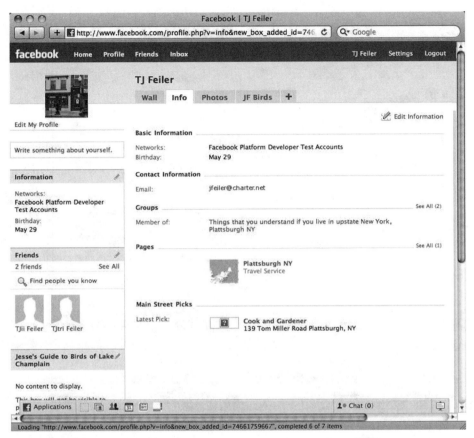

Figure 8.2

of the communication channels for messages automatically generated by your app. Even if a user decides to decline to pass a message along, no harm is done, and those that are posted can resonate with the user's friends.

But these are not all of the opportunities that Facebook can provide for your app. The name of the game is user involvement, and in the next chapter, you will find more ways to allow users to interact with one another.

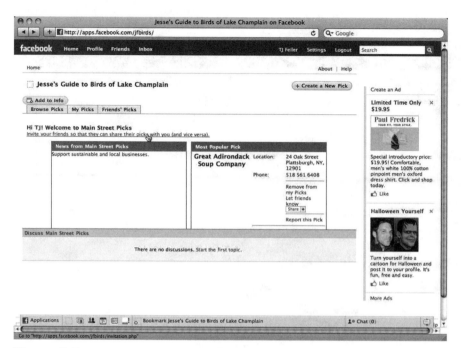

Figure 8.3

Figure 8.4

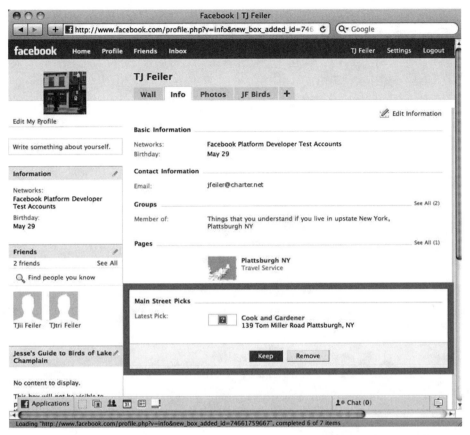

Figure 8.5

9

How Facebook App Integration Points Increase User Involvement

IN THE PREVIOUS CHAPTER, you saw how Facebook integration points work in general. This chapter delves further into what you can do for the user to heighten his or her involvement with your app. And remember that involvement normally leads to more time spent with the app, more uses of the app, and recommendations of the app to friends.

How Users Set Up Integration Points

Sometimes, when people start using Facebook, they work through and configure the various settings for Facebook options, but it appears that these people are the minority. Most people use the out-of-the-box settings (which is one reason why Facebook takes such pains to make the default settings appropriate). When you think about developing an app, you need to be aware of the wide variety of users and user settings. These settings and their defaults change over time as Facebook grows and expands in different areas. The settings also change as users add apps and as new features emerge. Part of building Facebook apps is keeping up with the changes in these settings so that you can provide the best experience for your app's users. You do not want to litter your app with instructions and warnings, but by bearing in mind that different people use apps in different ways, you can provide multiple methods for users to become involved with integration points. Only the most determined will use all of them, but having as many as possible available in as many ways as possible can increase the probability of users taking advantage of at least some of them.

After a user has added an app, it shows up in the list of apps in the Application Settings section of the Account link. The user can then click on an app and configure it with the settings shown in Figure 9.1. These

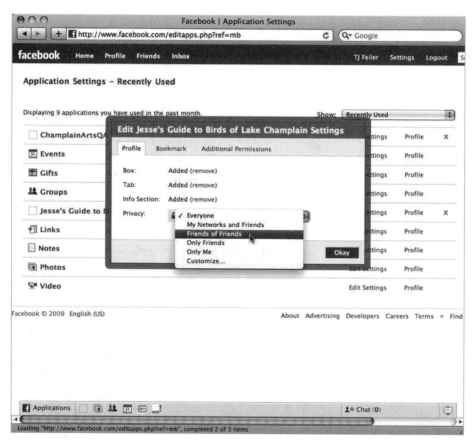

Figure 9.1

include privacy settings to control which groups of people (if any) can see the user's interaction with the given app. In addition, the user can specify whether the app can add boxes or tabs to the profile page and whether it can provide information on the Info tab of the profile page. This is a structured interface that is common to all apps, although not all of them offer all of the integration points. Remember to provide a variety of options in your app so that a user who might only know how to configure one type of integration point can configure it while leaving the others untouched.

You must provide integration functionality to even be added to anyone's list of apps; this is the first critical step for you and your app. Users will see that a particular integration point is available and be able to click an Add button. If that button is not available, that's the end of the process. And, as evident in Figure 9.1, having added an integration point for an app, a user can remove that integration point (and re-add it in the future).

From the user's point of view, this means he or she is in control. From your point of view, it means that if you do not provide these integration points, you have lost the possibility of providing further integration and user involvement with your app.

■■■ Add App Content to the User's Profile Page

You have learned how to add small amounts of information to the profile page in areas such as the feed. Here are more substantive additions your app can provide to profile pages at the user's request:

- Profile boxes in the narrow left-hand column or in a Boxes tab on the profile page
- An application tab specific to your application that provides a large area of space for your app's use

Both of these types of additions increase user involvement and the potential cascade effect of heightened involvement from friends.

An app can post a profile box to a profile page at the user's request in the Application Settings portion of the Account link (see Figure 9.1). Profile boxes can appear in either the left-hand column on the page beneath the user's photo, information, and friends, or in a Boxes tab on the main profile. The app specifies whether it will be added to the left-hand column or the Boxes tab on the profile page. If the app supports it, the user can move the box from one area to another by clicking the pencil icon at the top of the box. Although the box can be moved, it can appear in only one of the two specified places. Figure 9.2 shows a user's Boxes tab containing boxes from two separate apps.

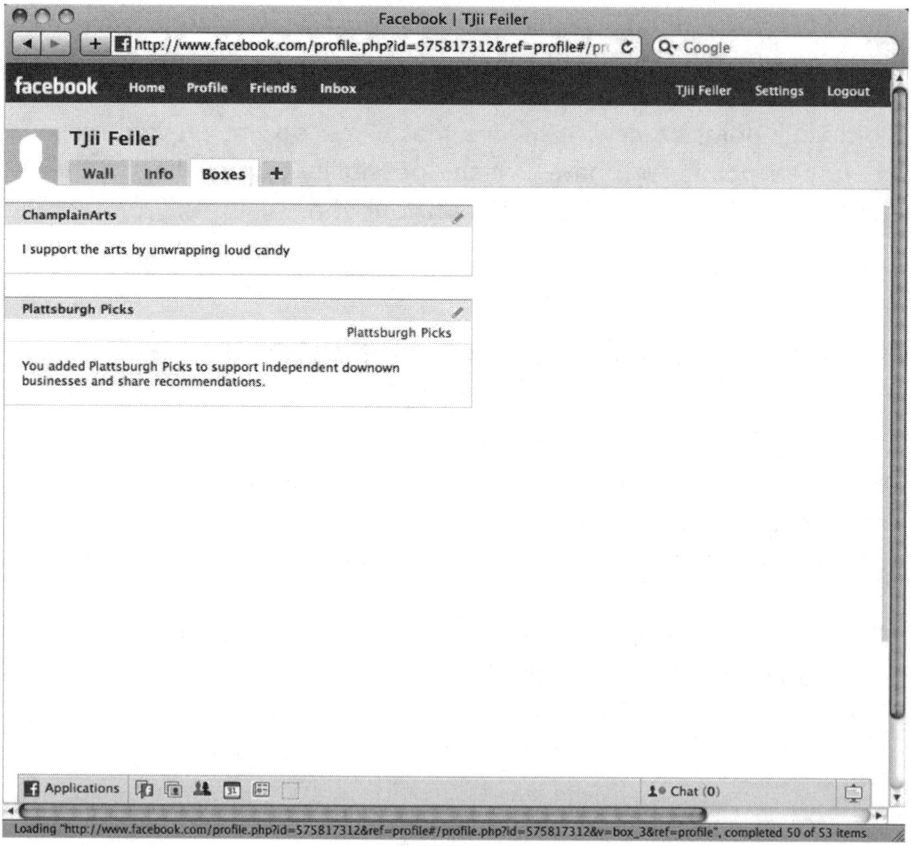

Figure 9.2

Avoid the temptation to tell users where to place the box, and in most cases, allow them the greatest flexibility possible. What a user does with your profile box is really a secondary (or even tertiary) concern for you: you want him or her to use the box and be involved with the app. If the user places all profile boxes for apps beginning with the letter *g* in the left-hand column, so be it. Facilitate user involvement but do not attempt to control it (although suggestions are usually welcome if they provide the user with a new insight or functionality).

This box should contain information about the user's most recent interaction with the app. Space is limited, but it is sufficient to inform the user's friends what your app has helped the user accomplish. Thus, your app's message spreads.

■ ■ APPLICATION TABS

The last tab on the right at the top of the profile page is a + tab. It lets users select from the apps they have added and allow any of those apps to have a tab on the profile page. The apps can post lengthier information than is possible in profile boxes or the Info tab. Figure 9.3 shows an app's tab in action. (The example shows text, but you can let your imagination soar as you generate information for your app's tab. Graphics, links, and other interactive elements are powerful tools to deploy, but remember to make them relevant to your app and your goals for the app.)

■ ■ ■ Requests and Invitations Increase User Interaction

Facebook provides app developers with tools to let users interact easily with their friends. Often this is for the purpose of inviting them to use an app, which in turn can lead to recommendations to other friends and repeat visits to the app. A button or link on

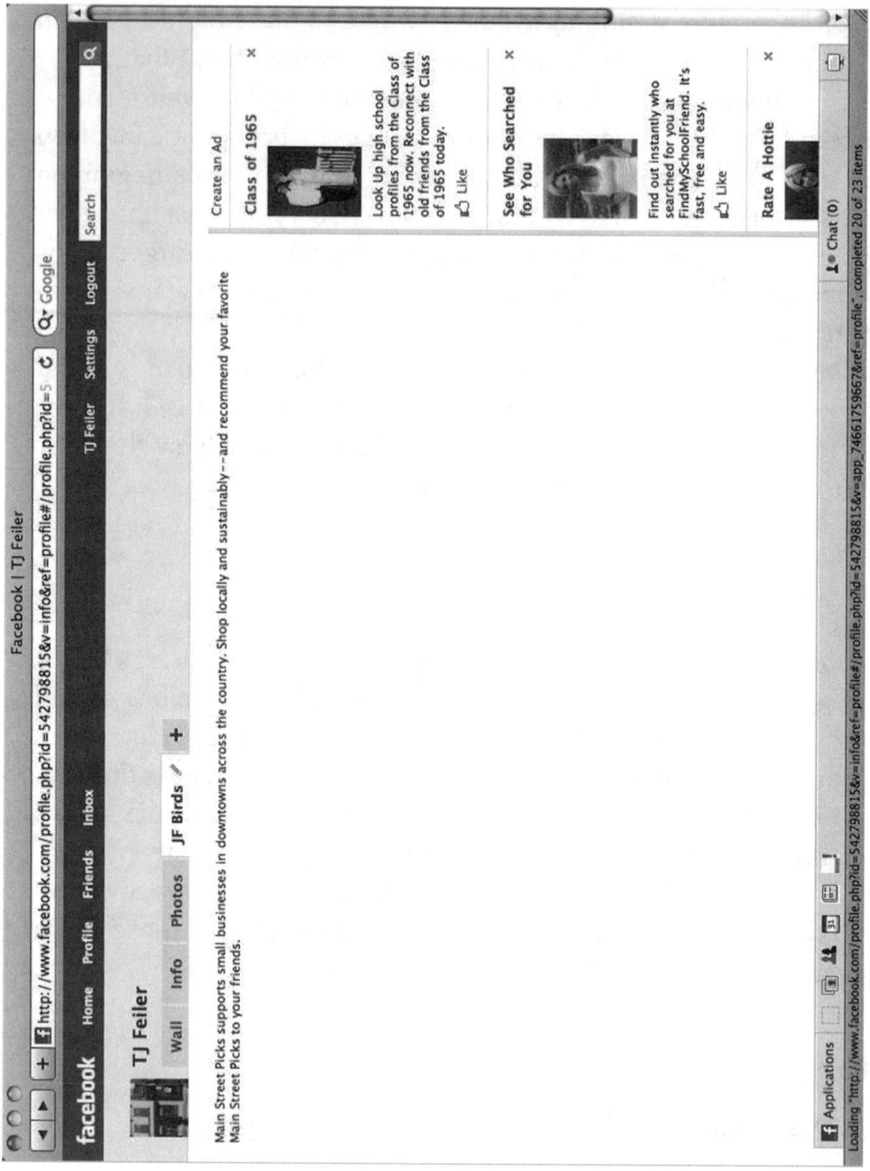
Facebook | TJ Feiler
http://www.facebook.com/profile.php?id=542798815&v=info&ref=profile#/profile.php?id=5
Google
facebook
Home
Profile
Friends
Inbox
TJ Feiler
Settings
Logout
Search
TJ Feiler
Wall
Info
Photos
JF Birds
Main Street Picks supports small businesses in downtowns across the country. Shop locally and sustainably--and recommend your favorite Main Street Picks to your friends.
Create an Ad
Class of 1965
Look Up high school profiles from the Class of 1965 now. Reconnect with old friends from the Class of 1965 today.
Like
See Who Searched for You
Find out instantly who searched for you at FindMySchoolFriend. It's fast, free and easy.
Like
Rate A Hottie
Applications
Chat (0)
Loading "http://www.facebook.com/profile.php?id=542798815&v=info&ref=profile#/profile.php?id=542798815&v=app_7466175966 78&ref=profile", completed 20 of 23 items

Figure 9.3

your app starts the process and brings up the window shown in Figure 9.4.

Users can select their friends based on photos (if they have uploaded them) or by typing in the names. They can also send e-mail invitations by typing addresses in the box at the bottom of the page; this means the invitations are not limited to Facebook users. Clicking the Send Invitation button at the lower left of the window brings up a preview of the invitation that will be sent, as shown in Figure 9.5.

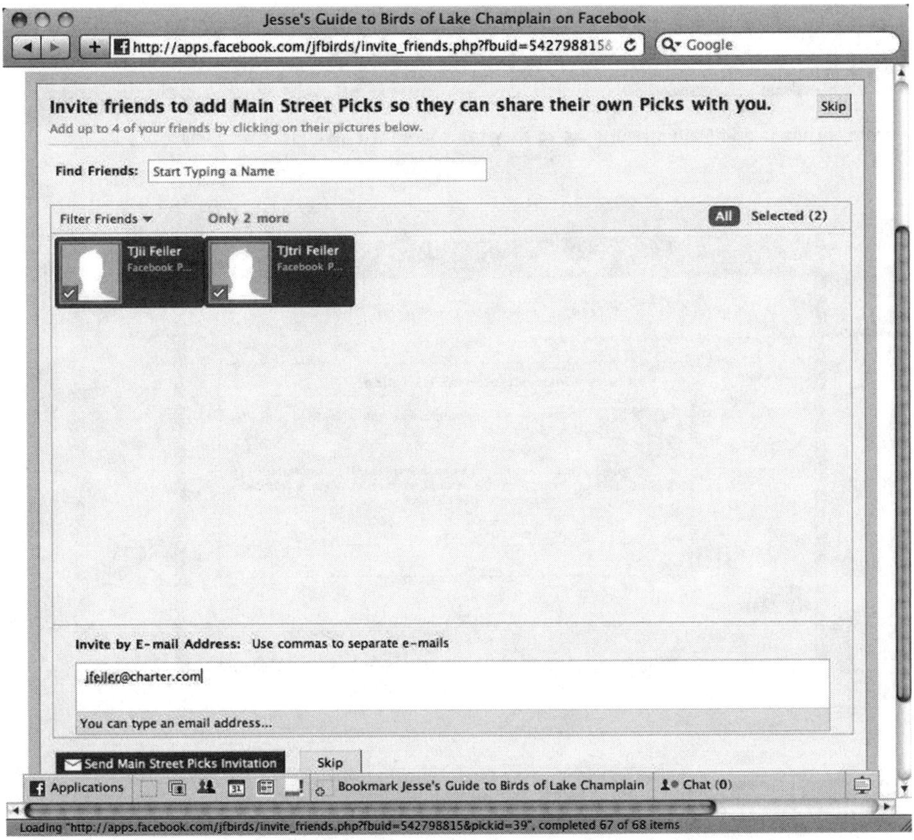

Figure 9.4

This is an incredibly powerful way of expanding your app's user base, but you have to be careful how you use it. Facebook's terms of use forbid you to exploit the feature, such as by requiring users to invite a certain number of friends to use the app before they themselves can do so. (This rule was put in place after some imaginative app developers came up with the initial use requirements gambit.) In addition, you are limited in the number of invitations your app can send in a given time period (notice the "Only 2 more" message at the top of Figure 9.4). This limit is dynamic and based on your app's track

Figure 9.5

record as measured by Facebook's internal monitoring; the developer discussion group always has spirited debates on this topic. Search on "invitations" to see the latest.

■■■ Using a Share Button

Users can also share content by using a Share button, as shown in Figure 9.6. Many people are used to such buttons, with labels such

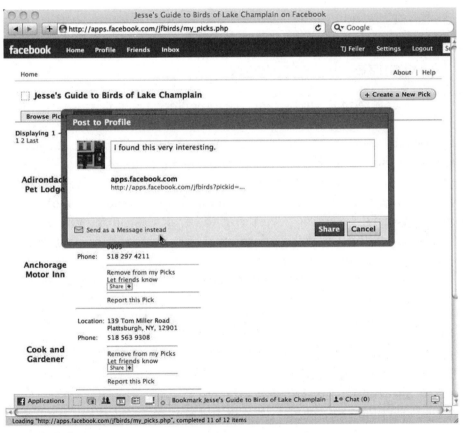

Figure 9.6

as "Share on Facebook" or "Digg It." Figure 9.6 is drawn from a demonstration app that lets people list local businesses and choose their favorites. In the background, you can see that, under the address information provided, users can add each business to or remove it from their list of favorites. They can also let friends know about a business—this link sends an invitation just as the invitation feature described earlier does.

When it comes to sharing information with a Share button, users can choose to post it on their profile or to send the information to specific friends or e-mail addresses.

■ ■ ■ Constructing Integration Point Content to Maximize User Involvement

We have covered the various integration points available to your app and, through it, to your users. These features make the user's interaction with your app more intense and rewarding. They also let you and your users leverage the social graph—the array of friends— on Facebook. It is this social network that is so important to people as well as to advertisers. A recommendation from a friend is one of the most powerful forms of advertising there is. Even a fairly low-key recommendation—just the information that someone is using your app—can often be compelling.

To make the most of integration points, you have two sets of tasks. You need to implement them in your app, and you need to implement them according to Facebook's rules and in the best way to further your goals for the app. Do you want the app to spread as quickly as possibly through your users' social webs? That might be the choice for an app that is tied to a time-specific event such as a movie release or a retail store's sale. On the other hand, your goals for the app might be to build a brand and brand loyalty; that might be best achieved with a slower spread but more intense (that is, more detailed) user involvement. You have already seen the basics of how to implement integration points (further details for the programmer are available in the online documentation in the Developers section of the Facebook site). This section provides suggestions about the best ways to implement integration points and how to structure the information.

■ ■ GETTING TO AN INTEGRATION POINT

The various figures in this book have shown buttons and links that lead to integration points. In Figure 9.6 you saw how a Share button can be used. That button is part of the Facebook tools for developers. The link "Let friends know," which appears above the button in each list entry, provides an entrance to the invitations discussed earlier. These are the two ways of getting to the integration points from your app: using Facebook-provided buttons or using links that you create for your own graphic or text.

As a general rule, if a Facebook button is available, you should use it. Users will recognize it and know how to use it and what it does. Furthermore, Facebook has already trained its users. The integration points are the heart of Facebook, and Facebook promotes them and provides information about them to users. In cases where there is no Facebook-provided button, Facebook provides routines for programmers to use to link their apps to the relevant functionality.

A critical part of your app's design is how you will use the integration points. The links and buttons should be right where the user's attention is focused so he or she knows exactly what will be shared or what an invitation will contain.

■ ■ STRUCTURING INTEGRATION POINT DATA

So far, you have seen integration points from the perspective of the person who is using the app. What goes on behind the scenes is clear only when you look at the results of the integration points. Figure 9.7 shows part of a user's feed with a variety of integration point stories, including those that resulted from using the Share button in Figure 9.6. This sequence provides a good example of how apps leverage the power of integration points and get exposure for your app. The feed story con-

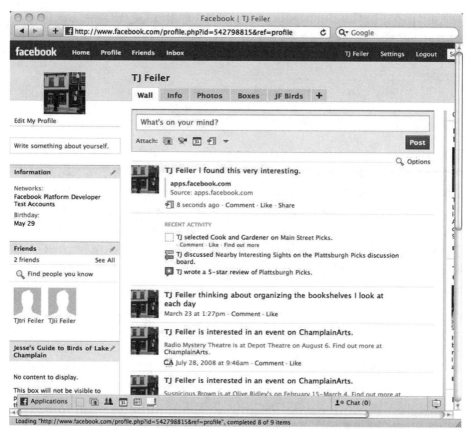

Figure 9.7

tains the text the user typed in; it also contains the user's name, a link to the item that is being shared, and the source of the item. Everything except the text the user typed in is automatically passed into the profile. The app is programmed to provide those links and the text. From both your standpoint and the standpoint of the user who is sharing, automating integration with the social web using one click is the most productive way to leverage Facebook's network of friends.

What is not immediately evident from the static image in Figure 9.7 is that it is riddled with links so the person looking at the feed can pursue the story. For example, Figure 9.7 is built on the demonstration app that lets people list businesses in their area and then select their favorites. Under Recent Activity, the first item is "TJ [the user] selected Cook and Gardener on Main Street Picks." Main Street Picks is a link; clicking on it takes the user to the app. If you program the app properly, this link will take the user to the specific section of the app devoted to that business. (The links are more visible when shown in Facebook because they are displayed in blue while the main part of the story is in black text.)

One way to make the most out of the feed entry is to sketch out what you want it to look like. Your app automatically generates feed entries when a user interacts with it (that is the point of the feed). Usually it is not too difficult to figure out where in the app such stories should be triggered; they should be generated based on user actions but only when those actions can somehow create an interesting story. Changing a background color in an app's preferences is normally not an interesting story unless the app is about colors. Stories in the feed can be more interesting and involving if they include links. Remember that you are limited in the length of your feed stories; they should not be long-winded, but brief and to the point. Try to make the links in the story self-evident, such as a link to the specific part of the app referenced in the story.

This is an example of the "just works" theory of user interface. When it just works the way the user expects it to, you do not need elaborate explanations. One way of making things just work is to structure the environment so the user's expectations are pointed strongly in a given direction. The wording of invitations should be concise, but you can certainly spare a few words (fewer than ten is best) to give an example of what will be shared and what the result will look like.

Integration points belong in the initial design of the app along with details about the feeds and the interaction between the user and the app. Particularly when you are building an app for a specific purpose such as selling, promoting products and services, or providing training, the more you can incorporate the integration points into the app itself, the user's expectations for the app, and your expectations of the project, the more successful you will be. You may not implement all the points, but you should know what they are and how they will work. Going back after the fact can be difficult and, more important, you may forget to do so.

By incorporating integration points, Facebook makes your work easier. Both iPhone and Facebook want your apps to succeed in their own right and in satisfying users: they need your apps to enrich their platforms. That is why they both give you so much information to help you improve your apps. Be creative in the substance of your app, but not in how you enable users to perform standard Facebook functions. The more "Facebooky" your app is, the more people will use it and feel comfortable using it. Remember: apps for Facebook and iPhone rarely come with instruction manuals. If your app needs a manual, it needs a rewrite.

In the case of feeds, Facebook has been moving into richer and more complex stories and has been supporting them with tools. The most important has been the move from totally free-format stories to highly structured ones. This started with the ability for app developers to create templates for their apps to use in stories such as: "<actor> selected <item> on Main Street Picks." At runtime when the user carries out the triggering action, the app calls code that passes in the variable information. <actor> is automatically filled in by Facebook with the user's name and a link to his or her profile. In this case, <item> is defined by the app, and at runtime the app is responsible for passing in Cook and Gardener and indicating that it is to be placed in <item>. Main Street Picks, which contains a link to your app (always include links to your app and the app name in the feed), needs no replacement because it is the same for every story.

After a year of using templates, Facebook simplified things to generalize the production of these stories; now it talks about updating the *stream* that contains this information. Templates are gone, and in their place are other structures that have slots your app fills in at runtime for links, captions, media, and the like. The Facebook experience is designed to be consistent whether users are running it on a traditional computer, on the iPhone Facebook app, or through Facebook Connect and your existing website. Now it is possible to update a user's news feed using the stream architecture in all of these environments.

It is worthwhile looking at streams to see the options Facebook provides. The technical documentation is only a few pages long, so if you are working with a developer, spend some time going over it to see what you can put in your stories. The basis is technical, but the concepts are not, and the goal from everyone's point of view is to create a stream of interesting stories that involve users in a specific way. How do you tell if a user is involved in a story? There's one simple test: does he or she click something? It may be a standard link such as Comment, Like, or Share, but it may also be a link in your story. That's the point of each story in a news feed: a user click.

A good test of your feed stories is to list them all on a piece of paper before they are programmed. Match each one to your app's objectives. A story should always further those objectives. Remember that the stories appear in the feed along with other stories from other apps and the content the user enters through the Publisher. Every user's feed is different, and each one reflects the interests and personality of the individual, so you cannot tell in exactly what context your story will appear. The most you can do is make it serve your app's objectives. Depending on those objectives and who your users are, you may be able to fine-tune the stories (adding or subtracting humor and cuteness, for example).

All of the features in iPhone and Facebook apps combine to provide a user with value—be it entertainment, information, or your unique combination of the two in your area of expertise. There is one additional ingredient you need to address: the overall rules for succeeding in the app world, which are covered in the next chapter.

10

How to Make Your App Successful

THE PRECEDING CHAPTERS HAVE shown you the features and techniques that users expect in apps; fitting into that pattern can help your app succeed (save your innovation and creativity for the content). Now it is time to explore the other points you should consider. These are less technical issues such as competition and the general aspects of market research.

Do Your Homework

Building apps can be a lot of fun; it is a great hobby if you are a hobbyist. It can also be a terrific way to learn modern programming techniques—particularly the design of Web-based applications, the use of JavaScript, and modern data storage mechanisms. If you want to make money from your apps (that is, let your apps make money for you), you have to treat the process seriously, and that means doing your homework about the marketplace, the competition, and the challenges your app will face from both other apps and other media and technologies that are not even apps.

◼ ◼ EXPLORING THE MARKETPLACE

You already learned in earlier chapters how to watch yourself and others as they use apps on iPhone and Facebook. Now it is time to bring you up to speed on the app marketplaces. Even if you do not intend for your app to be a moneymaking venture, you are in a marketplace. There are thousands of apps out there that people can use for free, and your app is competing with all of them, whether you charge money for it or not.

The techniques for exploring the app marketplace are the same as they are for any product in any marketplace—nature tours in the travel market, cakes in a bakery, or shirts in a clothing store. Knowing your marketplace means knowing its definition, its size, and as much about its constituents as possible.

When it comes to apps, there is a temptation to be "app-centric" For example, you might define your market as iPhone users who are interested in nature travel. Someone else might choose nature travelers who use an iPhone. There is an important difference in focus and emphasis between these two definitions: are people more interested in the technology or the content? Will they say, "Look what I can do with my iPhone," or will they say, "Look at this app about a nontropical rainforest"? In some cases, it does not matter. For example, if your market is Apple developers who use an iPhone, the two groups have great crossover.

Defining your market is one of the areas of app development that benefits greatly from brainstorming and questioning every assumption. You may be tempted to construct a checklist of questions to be answered to define your market. That can be a valuable tool, but when you have answered each question, do not put the checklist away. This is a fast-moving world, and you have to rethink and requestion each assumption you make. An app that is totally unrelated to the app you envision can emerge tomorrow, and it may suggest new ways of doing things to both you and your users.

■ ■ STUDYING THE COMPETITION

Once you have defined your marketplace, look to see what your competition is. If you have an idea for an app for nature travelers, look at others who have apps in this area. Also look at the competition in the broader market of general travelers. There are two reasons for doing this.

First of all, you need to know if there is a general travel app that could easily be repurposed for the nature traveler. If there is, it might be a day's work for the developer of the general travel app to do so, and you should think carefully before devoting your own time to the project. But if your idea cannot be done easily by others and fulfills an existing need (or a need that can be made to exist), you are on the right track.

■ ■ READING REVIEWS

Study reviews of existing apps. Cast a wide net, looking at formal reviews as well as buyers' comments on blogs and Internet sites. What do users like? What do they not like?

As you read what's been said about other apps, you may find a common pattern. Many apps are rated by their users, and here is what you will generally find:

- A relatively small user base
- A large number of very low ratings

If you read randomly chosen reviews and follow the media as well as advice from Apple and Facebook, you will find that the low ratings are often backed up by comments pinpointing the issues. The most common complaint is "doesn't work." "Too hard to use" is also quite common, but "doesn't work" seems to be the winner.

Make sure your app works.

■ ■ CHECKING OUT WHO'S WHO

As you continue with your research, determine who's who in the marketplace. Perhaps you are a recognized expert in a particular field (a *domain expert*) and you want to use an app to spread your expertise. If not, maybe you should search out such a person. You should know who the media leaders and the influential people in the marketplace are.

In today's world of blogs and social media, you may find it easy to get in touch with media leaders. Don't think of the mass-market stars but rather the people who may be known to only a few hundred people—the *right* few hundred people. Read their blogs and reviews to see where they think the technology is headed.

■ ■ KNOWING HOW THE APPS MARKETPLACE IS DIFFERENT FROM OTHER MARKETPLACES

Knowing the market is standard advice to people developing new products, but it takes on a different flavor in the world of apps. In part, this is because apps have no physical distribution mechanism. You buy iPhone apps from the iTunes App Store and download them to your iPad, iPhone, or iPod touch. You use Facebook apps through Facebook. There is no package and no shipping. And Facebook apps as well as a lot of iPhone apps are free. These are some of the factors that can take an app's user base from zero to thousands in no time. (Remember that many apps also go from zero to only a handful of users in many months or years. The absence of a physical distribution channel can mean that an app that is downloaded once a year can remain in stock for a number of years. Once it has been set up in an online store, there is essentially no cost for keeping it there.)

The marketplace is heavily influenced by and is a part of the Internet. Bloggers play a big role, as do websites both large and small. The market leaders are often people like you. Apps are part of the modern democratization of software development: you do not need a staff of experts to deliver a great product. Facebook today employs just over a thousand people, and it serves 350 million, half of whom use it daily.

■ ■ ■ What Your App Must Do for the Platform

The formal rules for Facebook and iPhone apps are spelled out in terms of service, terms of use, and other legal documents on their websites. To gain access to the iTunes App Store, you must register as a developer ($99 per year at the time of this writing). As part of the registration process, you agree to Apple's terms for developers. When you become a Facebook developer, there is no fee, but you still have to accept Facebook's terms. If you are not a registered Facebook developer, your app cannot be part of a Facebook page.

The terms of service and other legal terms change from time to time, and this section does not interpret them for you. That is a job for you and, if the stakes are high, your legal advisor. However, it is simple to summarize what both Apple and Facebook are doing in their terms of service: they are protecting their platforms and their investment in those platforms.

Both companies have promised their users and customers that they will provide certain services. Their offerings are made richer by the many third-party apps that exist. You need Facebook or Apple (or both), but they also need you. Anything your app does to bring disrepute to Apple or Facebook is going to be (at the least) a violation of the spirit of the legal terms you have accepted. As noted previously, the most serious thing your app can do to damage Apple or Facebook is to not work. Apparently this cannot be repeated enough, because not working remains the major reason for both Apple and Facebook rejecting apps. Both have review processes for new apps that include running them. In some cases, this review process is the first time anyone other than the developer or the developer's immediate family has run the app.

One key step in development is finding someone you do not know to test your app. Find a person with some interest in the app's subject area and ask him or her to ask a friend you don't know to try the app. While you are at it, try out the listing you will put in the directory. Maybe it is something such as, "This app helps you pick the right variety of tuberous begonia to plant and the right date to plant it by using your current location on iPhone 3.0." It is not unheard-of for developers to go through several iterations of a blurb and an app until they get them both right. A number of developers have reported anecdotally that after the initial development of their app, most of the changes have been improvements to simplify and focus the app.

After development and refinement of the app, it is ready for submission. Note that you submit Facebook apps to be listed in the Application Directory. In addition, you can apply to have an app verified by Facebook. There is a charge for this (currently, $375 a year, or $175 a year for students and nonprofits), but it can be worthwhile if you are serious about building your app's credibility. Verified apps are listed before others in the Application Directory. For iPhone OS apps, the submission process happens through the developer program as described in Chapter 2. There is no directory or submission procedure for iPhone Web apps.

■■■ Facebook's Privacy Rules

Although Facebook looks for adherence to all of its terms, rules, and guidelines, the key issue on which it focuses is privacy. Facebook's databases contain a vast amount of data from its users, and the site has made promises to those users (and sometimes to governments) about how that data will be safeguarded and used. While your app is running, it has access to users' data, but you cannot store anything except meaningless ID numbers in your own database. You can store the ID number of an event or a user, but the name of the event or the user must always be retrieved from Facebook when you need it. This means that if the information has changed, you get the latest data. It also means that if the user's privacy settings have changed, you may not be able to get the underlying data for a given ID, even though you got it legitimately yesterday—or an hour ago.

Privacy includes protecting users from unsolicited contact. Facebook allows users to set controls for who can access their data and send them messages. If you violate those controls, you violate the user's contract with Facebook.

Violating Facebook's privacy rules (including restrictions on certain content visible to underage users) is serious for all concerned. Remember: Facebook has one trump card available. It can prevent your app from running on its platform if you go too far.

■■■ iPhone's Network Rules

iPhone has a different primary concern: it needs to keep communication channels open. If an app compromises telephony or the built-in networking functionality, it is not welcome on iPhone. The iTunes App Store is just over a year old at this writing, and everyone—including Apple—is finding out what the right mix of rules is. If you choose to push the envelope, the company may push back.

Apple has also made it clear that it does not welcome apps that compete with the built-in iPhone apps. You are invited to use the underlying APIs for access to data, but do not reinvent a wheel that Apple has already invented.

■■■ Joining the World of Apps 3.0

In Part 1 of this book, you have found an introduction to the exciting world you are joining. Maybe you are joining it for the first time in any capacity, or maybe you are a long-time app user who is just starting to think about making your own contribution.

Now that you have looked at the basics of apps and the specifics of iPhone and Facebook apps, it is time to move on and plan how you can use this knowledge and these technologies to get rich. The apps structure is a new way of developing and using software, so now is a good time to think about how to make the most of it. Part 2 will give you specific strategies for succeeding with apps.

Strategies for How to Get Rich with Apps

11

Strategy #1: Give Your Knowledge Away for Free—and Make Money!

FACEBOOK APPS ARE FREE, as are many iPhone apps. So if they are given away, how can you make money? There is a long history of this ranging from free maps at service stations in the 1950s to free broadcasts on radio and TV networks. Call it "being a loss leader," "promoting the brand," or "building goodwill"—your gift today can come back tomorrow in profits.

Apps offer three potential direct revenue streams:

- You can sell your app to users.
- You can sell advertising on your app.
- You can sell products and services through your app.

Facebook apps are free, so the first option is not available there, and it is not always available on iPhone because so many of those apps are free as well. This chapter explores the pros and cons of giving away your app. The remaining chapters in this part of the book explore ways to use the other two options. By implementing these strategies, you will be ahead of the curve.

Apps have joined the time-honored tradition of giveaways. All of the sales and marketing techniques that have been honed over the years for free trials, wine tastings, and imprinted items (such as key chains, sponges, pencils, and dog dishes) apply to apps. In addition, there are some new issues that apply only to apps.

Here's how to get started.

■■■Leverage Social Networks to Get the Ball Rolling

The numbers are enormous in the world of apps, with hundreds of millions of Facebook users and equally impressive numbers of iPhone users. Apps for Facebook and iPhone are counted in the scores of thousands. This means the opportunities are vast, but so are the challenges. You have a major ally in the viral nature of the social networking world. Do not just plan to give your app away; plan from the beginning to give it away to people who will then pass it along to their friends.

■ ■ GET YOUR APP INTO THE VIRAL NETWORK

You will give your app or your knowledge away for free, but users will pay you in one (or both) of two ways:

- They perform added distribution to their friends.
- They come back to purchase other products from you.

This means you have to make it easy for people to pass your free app along. Because iPhone apps must be downloaded and installed through Apple's iTunes App Store, you may want to consider an iPhone Web app instead of an iPhone app as a first step.

■ ■ WHAT NOT TO DO

There is one major area you should address right at the start. Remember to read your agreements with Facebook and Apple (as well as any other app players you are working with). Pay particular attention to the constraints you have agreed to regarding the privacy and sharing of user information. It is great if someone who is using your giveaway app tells a friend about it, and you have no control over the actions of your users (although there are some areas where you must exercise due diligence). But if you design your app to pass information along to the user's friends or contacts, and if that action is done by the app without the user's involvement, you are almost always violating the terms of service.

As mentioned in Chapter 9, it was common for early Facebook apps to set up a form of tollbooth, whereby users could not enter an app until they had invited a certain number of their friends to use it. Although technically the users were inviting their friends, the fact that the app enabled such invitations (a good thing) was considered inappropriate because its functionality could only be unlocked by forcing the user to invite friends (a bad thing—notice the word *forcing*). Remember how easy it is for users to give up on an app they do not like or find irritating. The one-click ease of launching an app is comparable to the one-click ease of closing an app's window forever.

■ ■ MAKE YOUR APP VALUABLE

Particularly with a free app, you want to deliver value so people will use the app and share it. Remember that value is in the eye of the beholder (but it starts in your brain when you think it up). With apps, much of the value comes from their presentation—everything from the name to the graphics. You have a small window of time to grab a potential user's attention and present your app in such a way that he or she can decide whether or not to try it. A major part of the presentation must be positioning it in such a way that potential users understand what it can do. The greatest app in the world will not be successful if a user downloads it thinking it is something it is not.

And no matter what the opportunity or goal for your app, there are some basic guidelines for creating free apps for either Facebook or the iPhone. You will not go wrong if you apply them to any app, but they are critically important for free apps.

■ ■ ADD SOME WOW TO YOUR APP

What you are looking for is a "wow factor": an app that is so cool or so useful or so new that people react both by saying, "Wow!" and telling their friends. A private wow is not much use to you. The last thing you want is for someone to try your free app and throw it away saying, "You get what you pay for."

Sometimes wow is fleeting—a fad. Because Facebook apps and iPhone Web apps are so easy to access, they are slightly more likely to become fads than iPhone apps are. There is nothing wrong with a fad, as long as you can recognize it as such and plan accordingly.

One of the advantages the digital world has over the physical one is that products are invisible: they do not have to be stored, shipped, or packaged in the traditional sense; all of this happens electronically over the Internet. Once an app has been created, the incremental cost of selling each copy of that app is almost nothing. This completely avoids the problems of fads in the real world: warehouses full of last year's fad, not to mention the very real possibility that by the time someone gets around to shopping for a fad product it may be out of stock or even out of fashion. The digital world of apps is essentially infinite and immediate. You cannot run out of stock in most cases. This means you can take a chance on something that is (or might be) a fad.

There is one important exception to this that can trip you up if your app takes off like a rocket: your server infrastructure. This comes into play in different ways for different apps, as you will see later in this chapter.

Free apps often take off more rapidly than apps with even a modest cost. Giving away an app for free does nothing for your bottom line unless lots of people use it, pass it along, and then possibly come back for additional paid services. Because you may have sudden spikes in users, you need to prepare a plan for dealing with that traffic.

Many iPhone apps are self-contained. You write the app, test it (and retest it), and then pass it on to the App Store. Once it is accepted, it is listed in the store's inventory, and people can download it directly to their iPhones. Whether one person or a million download it, the load is on the App Store's servers, which are designed to handle enormous amounts of traffic.

But not all iPhone apps are self-contained in this way. Many of them integrate the Web, which is actually one of the most important features you can add to your app. If the Web integration involves accessing a website you manage, the difference between one user and a million hits home: right in your Web server. It has to be scalable on extremely short notice to accommodate sudden popularity. It also has to be equally scalable on the downside if and when the next app comes along and your traffic drops suddenly.

The days of glowing media reports about an app or a website that became so popular that it crashed are long gone. Overload is no longer a sign of success: it is a sign of poor planning and failure.

How can you prepare for possible spikes? If your Web infrastructure is already big and scalable, you are home free. But if you have a small business with a website that may be hosted with a shared hosting package, you can plan for contingencies.

This issue is becoming a common one, and there are services designed specifically to handle quick scalability, including companies with large Web infrastructures such as Google and Amazon. These infrastructures already have load balancing built in, and they are key players along with others in the new architecture called *cloud computing* (there is more about this in Chapter 17). They provide hosting services at reasonable costs to generate additional revenue for their primary businesses (providing search results and selling merchandise). Because their data infrastructures are already so big, they can usually accommodate a quickly upscaled or downscaled application.

On iPhone, this is a potential problem only if your app needs access to your server during use. It is a much more common problem for Facebook apps, which always require access to your server as they run. That is why you may want to explore the developer section of Facebook to find hosting services. They will be happy to host your iPhone support as well as your Facebook app.

Managing the Flood of Users with iPhone Web Apps. iPhone Web apps may be the most vulnerable to a sudden flood of users. The iPhone Web app runs in a browser on a user's iPhone, and that browser communicates with your Web server. There is no App Store or Facebook infrastructure to provide part of the user experience. It is all provided by your Web server.

The exceptions are functions that Safari can perform on iPhone. Because those features can add tremendous value to your app, they can provide you with a double benefit: added value that does not impact your Web server.

Handling the Volume with Facebook Apps. Facebook apps are the most complex when it comes to performance. Facebook itself provides the log-in infrastructure and the frame of the page, while your Web server provides the content in the center of the page. Everything needs to work together properly to deliver the app to your users.

However, that complexity pays off mightily when it comes to handling spikes in usage. The reason is the Web server that provides the content of the Facebook page is your Web server only in the sense that you control it (or have access to it, in the case of a shared server). The URLs for your app's pages are Facebook URLs; your server's address is not visible to users. This means you can easily host your Facebook app somewhere other than on your main server (and this is what many people do). You will find a number of hosting services listed in the developer section of Facebook and can take your pick.

You may use this type of architecture for an iPhone Web app as well: you can off-load the support for the app to another server that can handle the load more easily than your main server might be able to do.

■ ■ Keep It Simple

In addition to the wow factor, you should be looking for simplicity. One way to do this is to start by writing the brief description of your app that will appear in the iTunes App Store, as shown in Figure 11.1; the list of Web apps on Apple's site (apple.com/webapps), as shown in Figure 11.2; and/or the Facebook Application Directory (facebook .com/apps/directory.php), as shown in Figure 11.3. Start writing these descriptions as soon as you start thinking about your app; it takes time to refine them.

Browse these listings to see how they are constructed as well as which ones catch your eye. You can also check out the developer resources for Facebook and iPhone for tips on constructing good descriptions. Remember that Facebook and Apple rely on third-party apps, so they want to help you write the most effective descriptions possible.

Innovation is terrific and creativity is priceless, but they are not always good ideas. The app listings are all divided into categories. It is true that some major marketing successes have come from products that invented their own categories, but that can be a risky and expensive way of launching a product. The categories in the app directories can evolve over time, but from the moment you start thinking about your app, make certain you know where you will place it in those directories. That can be the beginning of making your app's message simple. This is particularly important with free apps, because you normally want to cast the widest possible net for new users.

Although creating a new category is difficult, it can be a success if a product redefines an existing category and implements functionality in new ways within that category. iPhone was not the first smartphone, and Facebook was not the first social networking site, but both products provided a new perspective on existing product

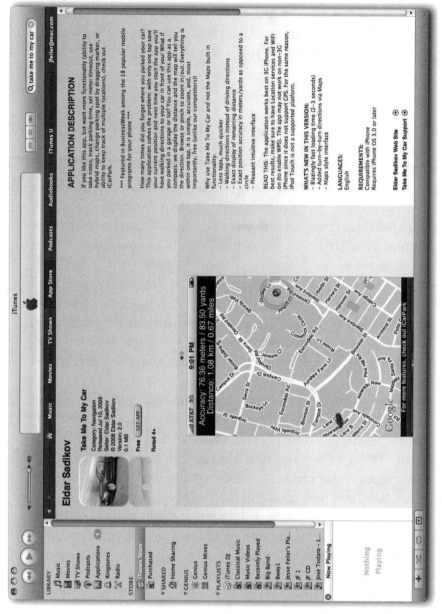

Figure 11.1

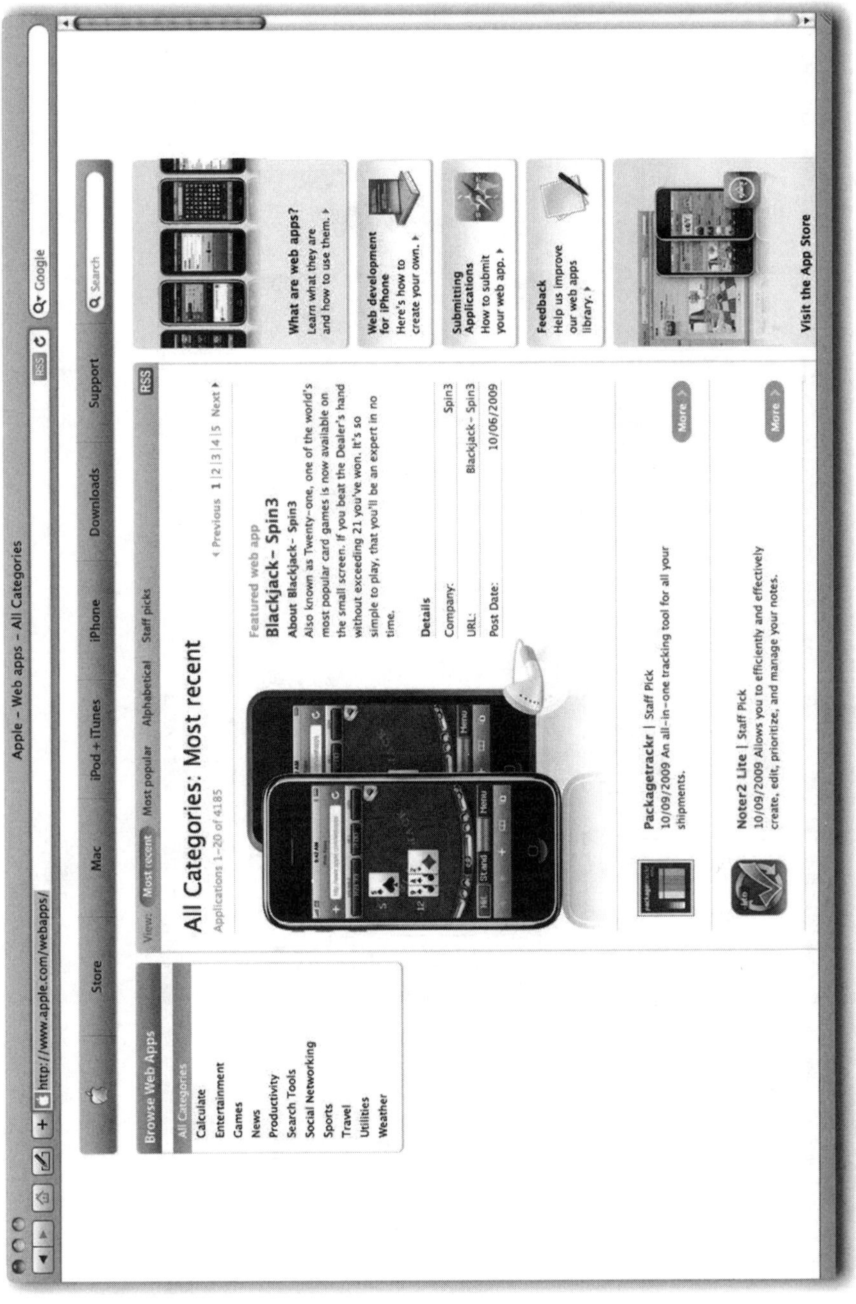
Apple – Web apps – All Categories
http://www.apple.com/webapps/
Store
Mac
iPod + iTunes
iPhone
Downloads
Support
Browse Web Apps
All Categories
Calculate
Entertainment
Games
News
Productivity
Search Tools
Social Networking
Sports
Travel
Utilities
Weather
View: Most recent Most popular Alphabetical Staff picks
All Categories: Most recent
Applications 1–20 of 4185
Previous 1 2 3 4 5 Next
Featured web app
Blackjack – Spin3
About Blackjack – Spin3
Also known as Twenty-one, one of the world's most popular card games is now available on the small screen. If you beat the Dealer's hand without exceeding 21 you've won. It's so simple to play, that you'll be an expert in no time.
Details
Company: Spin3
URL: Blackjack – Spin3
Post Date: 10/06/2009
Packagetrackr | Staff Pick
10/09/2009 An all-in-one tracking tool for all your shipments.
More
Noter2 Lite | Staff Pick
10/09/2009 Allows you to efficiently and effectively create, edit, prioritize, and manage your notes.
More
What are web apps?
Learn what they are and how to use them.
Web development for iPhone
Here's how to create your own.
Submitting Applications
How to submit your web app.
Feedback
Help us improve our web apps library.
Visit the App Store

Figure 11.2

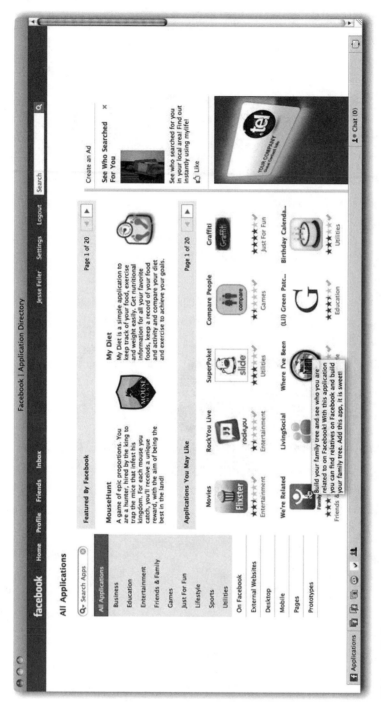

Figure 11.3

categories. Apple and Facebook thought they could understand and communicate to consumers more effectively than the product category pioneers.

Many people who have developed software in other environments have trouble developing apps, and often that trouble comes from adding too many features and cluttering the app. Consumers have been complaining about "feature creep" in desktop applications for years—applications becoming bloated with added features piled on top of one another in each succeeding version. You can look at the world of apps as a partial reaction to feature creep in the apps' small size and focused functionality. More is not always more; it can just be confusing.

Justify every feature and function you consider adding to your app, then go back and justify it again. Keep the app simple and be able to describe what it does (or what it helps users do) in as few words as possible.

This does not mean that your application has to be bland and generic. Know what it does and who its audience is. Markets for apps are often enormous, but because you are dealing with an Internet-based marketplace, that enormous market can be spread over the entire globe. An app that is helpful to bakers can be simple for them even if it is mystifying to plumbers.

Now that you have started thinking about the app you will give away, here are some ideas about the kind of app it can be.

▪▪▪ Offer "Lite" Versions for Free

If you are trying to promote an app that you will be selling, you can offer a free version for people to try. It may work for a limited time, or it may be limited to a certain amount of data.

The experience with a free version has to be positive enough to bring people back to buy the full version. They must see the power and functionality of the full version even in a limited trial. Pay particular attention to any feature that you change for the trial version, because you will be encouraging people to come back and buy a version that implements that particular functionality differently. (It is not uncommon for people to complain that a trial version is easier to use than the full-price version; that is an example of feature creep in action in the full-price version.)

Do not think that try-before-you-buy is limited to selling apps. You may be selling a complex engineering product for networked personal computers and implement one specific feature as an app. In a case such as this, the try-before-you-buy app presents one feature of a much more complex product as an app. If the user wants to upgrade to the full product, it may run on personal computers or even a corporate network. The app that people try out can present your expertise or the functionality of the full system in a self-contained format.

■■■■ Free Apps Can Market *Anything*

You can use a free app to help you market just about anything. Building an iPhone Web app that accesses a specific RSS news feed is a matter of a few minutes' work. You can easily generate such a news feed from a blog (it is a built-in option for most blogging software). Make it a point to update that blog on a regular basis. For a development project that unfolds over time, such a blog-based RSS feed can provide behind-the-scenes glimpses of the project and build interest.

In the case of a movie or TV show, for example, someone can be designated to blog fifty words a day about what is happening on the set. Pick up those postings in an iPhone Web app or a Facebook app, and you have a channel with which to promote interest in the project. Obviously people can subscribe to a news feed from their own computers, but encapsulating it in an iPhone Web app or a Facebook app makes it more specific. And unlike a generic news reader, these apps pick up your news feed and only your news feed.

You can leverage any events in a blog that is picked up in an app. Think about a countdown to a trade show, and don't forget opportunities for user involvement such as inviting people to suggest seminars and meetings.

If you use an app that picks up a blog, one decision you have to make is whether to end it or not. For a movie, such promotions often end with the premiere. For a trade show, you might want to increase the pace of blog entries for your app as the show approaches. When it closes, you might send a farewell message or consider going into a lower-volume mode of perhaps one message a month. This can keep the communication channel open until the following year's event.

One particular example of this scaling down is found in the world of politics. The pace of messages using new media picks up as an election approaches. The really sophisticated campaigns do periodic follow-up messages after the election.

For a lengthy project, you can also use blogs and news feeds to compress time. It is a little more work, but not prohibitively so. Create one posting a week for the entire life of your project—perhaps two or three years—but do not publish these postings. Then, when you get closer to the debut of the project, go back and publish one of the weekly postings each day, culminating with the completion of the project.

▪▪▪▪ Build Your Brand

You can make a vast amount of information available on an app. If you provide a compelling and useful interface to all that information, your app will provide a valuable service to people and build your brand at the same time.

A free recipe app can establish your bona fides in whatever section of the food world you choose. Likewise, a Facebook app that provides stain removal tips for clothing can establish your credentials in the laundry world.

You may want to experiment with apps that display RSS feed data or other easily created updates. You can build an iPhone Web app for an RSS feed in minutes; with Facebook, you can import the same feed to a Facebook page with a few clicks of a mouse. Wrapping a Facebook app around a feed is also a simple matter. All you need to do is remember to update the feed regularly: the cost is minimal, so giving it away can provide value to the user without breaking your bank. Experiment so that you have the basics of updating the feed down pat and can deliver on your promise of frequent updates.

The low cost of developing many apps is critical to this strategy. You can develop targeted, useful apps quickly and easily. As an example, take a small store that sells artisanal cheese. Would an app help in marketing the cheese? Such a store may have regular customers, but many others live far away and come in while they are on vacation. Once they have visited the physical store, let them use a Facebook or iPhone app to browse an online store and order more cheese.

Special circumstances lend themselves to the development of apps because you can target them so precisely. Imagine this cheese store is located near the northern U.S. border. On one side of the border, people use Canadian dollars and weigh cheese using the metric system. On the other side, people use U.S. dollars and weigh cheese in pounds and ounces. You can develop an iPhone app that provides weekly updates of prices and automatically converts currencies and weights.

You may think this is a very specific example, and it is. That's just the point. Because apps are so relatively cheap to develop, you can build one for such a very specific case and give it away. You would not have to convert too many tourists to repeat online shoppers to pay for this minimal investment. Do not just give something away: give away information that is relevant to building and defining your brand.

You can develop a lucrative business by giving your app away. With this strategy, you have seen how you can encourage sales with a free app. But there is another way to give the app away and make money: selling advertising on your app, which is Strategy #2.

■ ■ ■ ■ ■ ■ ■ ■ ■ ■ ■ ■ ■ ■ ■ ■ ■

12

Strategy #2:
Publish Ads on Your iPhone
and Facebook Apps

LET'S TAKE A LOOK at how you can make money from app advertisements. Like Strategy #1, this involves giving your app away for free; the difference is that with advertising, you are paid for your work by advertisers who buy space on the app. This has worked for many years in many markets, perhaps most notably broadcast radio and television in the United States. Hybrid models also exist; for example, magazines and cable television make money from both subscriptions and advertising.

If you are going to publish ads on your app, you need to know what advertisers are looking for and how you can contact them. This second point is simple with apps: most ads are placed through ad networks that bring advertisers and app owners together. This is a highly automated process, but you need to know the basics behind it. You will find that information in this chapter. You will also see how you can handle advertising on your own without going through a network. This makes sense in a number of cases, such as confined local markets.

A few years ago, there were no iPhone or Facebook apps (in fact, there were no iPhones or Facebook). Such apps have become a major source of revenue for their owners in some cases, and advertisers are now using ads aggressively on apps because these ads bring in results and the economics are favorable. Given that this area is growing so rapidly, the basics described here will most likely remain constant, but the players may change. For example, Google has continued to increase its advertising offerings to target apps more aggressively.

In the world of ad apps, big players such as Google and many advertisers can write copy, create visuals, and come up with special app ad–based specials. There is one thing they cannot do: find potential clients. They need you to deliver the app users who will view and act on the ads. To put it bluntly, you (the app owner) are asked to deliver eyeballs.

Here is how to do that and make money at it.

■■■ Where Is the Advertising Money?

Advertisers pay to publish their ads on media ranging from newspapers, broadcasts, and billboards to Web pages and now apps. If you search the Web to find out how much money is involved, you will find a variety of answers, but a cluster of estimates for the United States is in the low $400 billion a year range. eMarketer estimated that mobile advertising in 2009 would amount to about $416 million, and most of that would be on apps of one kind or another. The same September 2009 report estimates that this number will grow to $593 million in 2010, $830 million in 2011, $1.14 billion in 2012, and $1.56 billion in 2012. In late 2009, Google purchased the AdMob mobile ad network, which sells space on apps; the price was $750 million. All research estimates show similar increases in the money spent on advertising in nonmobile Facebook apps. This is a major opportunity if you know how to use it.

Many books and seminars (and even university degree programs) go into greater depth on this subject, but you can start to evaluate your advertising opportunities by looking at the advertising world in general. Here are two basic, yet critical principles to consider as you think about monetizing apps with advertisements:

- Mass marketing and direct marketing are different.
- Advertising can have cumulative effects.

Both of these are particularly relevant to the world of apps, in part because the delivery of app ads is highly automated, and thanks to that automation, accurate statistics on ad exposure can be collected and analyzed. This section shows you how and why this matters.

■ ■ MASS MARKETING AND DIRECT MARKETING ARE DIFFERENT

In a nutshell, mass marketing is undifferentiated advertising aimed at whoever happens to see it—skywriting is commonly used as an example. Direct marketing is targeted to specific people in one way or another: viewers of a certain TV show, owners of a specific brand or product, and so forth. It's all about who sees the ads.

Advertisers pay to put their messages in front of people, but from their point of view, not all people are created equal. One of the first steps in building an ad campaign is defining the *target audience*, the people you want to reach. There is no point advertising dog biscuits to people who do not have dogs.

You need to think about two target audiences when you are dealing with apps: the target audience the advertiser wants to reach, and the target audience your app wants to reach. The more they overlap (or are identical), the more effective your app can be for an advertiser—and the higher your rates can be.

On the other hand, a relationship that is not a total overlap can still be helpful. A dating app might be a good location for ads promoting everything from mouthwash to gym memberships and clothing. The one thing that is not going to be useful is a conflict between your app's target market and the advertiser's; you would not advertise a steak house on an app promoting vegan living.

Understanding Impressions and CPM: Mass Marketing Through Apps. The cost of advertising is dependent on the number of people an ad will reach. Advertisers often talk of impressions—each impression being one person seeing the ad—and the cost per thousand impressions (CPM). This terminology predates apps and even the Internet; it is how advertising has been discussed for decades. The CPM is so important to the economics of advertising that a lot of money is spent measuring how many people see ads. Television ratings, for example, are measured so network and cable stations can charge advertisers appropriately for certain time slots. When it comes to apps, measurement is automatic, and advertisers often have much better data with apps than they do with traditional media, in many cases because the automation of the app delivery channel can provide detailed feedback.

This data is available to you as an app developer, promoter, or owner. Look at your app's statistics as provided by ad networks (there is more on this later in the chapter). You can get an accurate idea of the number of users you have as well as honing in on who they are. That, in turn, can help you promote your app to the likeliest users as well as to an audience that will be most valuable to your advertisers.

Mass-marketing campaigns target large but relatively broad audiences. A local newspaper's sports section may deliver an audience that has a somewhat higher male readership than the same paper's decorating section, but that is about as far as many media go. The recent proliferation of cable stations has allowed television to create channels geared to people with specific interests, and that is good for advertisers as well as the stations themselves. Thus, although the total audience for a targeted show or publication may be low, the revenue and CPM may be higher than those for a mass-market show or publication.

The broader your app's appeal, the larger the number of users you can attract. If you sell ads on your app, you can deliver many people to your advertisers. However, an app with broad appeal may be able to charge only a relatively low CPM, so do not let those thousands or hundreds of thousands of users start you thinking about retirement.

If your app has a middling number of users and a broad appeal, chances are your advertising revenue will not be significant. By focusing your app more precisely, you may be able to get advertisers to pay more. Alternatively, you may decide to expand your audience so you can make money from a larger (but less precise) demographic.

Understanding Attention and Clicks. Until the Web came along, a lot of the information about ad audiences was hypothetical. How many people actually see an ad in a newspaper or on television is a matter of much study. The publisher can report how many papers were delivered, but that does not let advertisers know how many people actually saw the page on which their ad was printed, and they certainly do not know how many of the people who saw that page did (or did not) see their ad.

Television ratings services can estimate how many people saw a particular show, but they usually cannot always calculate how many people turned away during a commercial. Surveys can hone in on who readers and viewers are so that advertisers can see how a typical audience compares to their target audience.

This time-honored process for evaluating advertising has totally changed with the Web. You do not have to extrapolate from the number of papers printed to figure out how many people saw a specific page; you can look at your Web server log to see exactly how many times your banner ad was served up to users. That is not an estimate or extrapolation.

When it comes to the matter of attention, things are also different on the Web. Many Web ads are not simply displays; they include links. If someone clicks on your ad, you have the answer to the question of attention. If an ad is placed in front of someone, that person may or may not look at it. Likewise, if someone is watching a TV show, his or her attention can wander to something else. But in Web-based advertising, if a person clicks on an ad, you know he or she has seen it. Just as with the number of impressions, this number is not an estimate or extrapolation. Thus, in addition to CPM, ads on the Web and in apps can be characterized by a cost per click (CPC). Note that CPM is cost per *thousand* impressions, but CPC is cost per *single* click.

All of this ties into the comments about the rapid growth of the app ad market made at the beginning of this chapter. As the market grows, and as more and bigger traditional advertisers jump in, their demands for hard data cannot be ignored. In the olden days (a few years ago) when advertisers had to be told what Facebook was, they might have thrown a few hundred or even a few thousand dollars at a Facebook app just to see what happened. Now they want to treat ads on apps the same way they treat ads on all other media, at least when it comes to market and audience research.

The collection of statistics is automatic, but you have to play your role in understanding what the advertisers are looking at—and you have to look at it first. After your app has launched, if you want to make serious money from it, track your users not just by raw numbers, but by whatever demographic information you can find. In this regard, you can look in advertising textbooks for the best ways of measuring who your audience is. Polls in apps are popular with many users, but they can be even more popular with advertisers, particularly if you tighten them up to prevent double voting.

Deciphering eCPM. Advertisers shop around for apps that serve their purposes. For advertising campaigns that are spread over various media, being able to compare CPMs for all media can be useful, and that appears to have been the genesis of eCPM, or the effective cost per thousand, for your website or app. Some people believe this ratio is meaningless; others do not go that far, thinking it is harmless. Because you are in a competitive market when you start to sell ad space on your app, you should do your research so you know what the competition is charging and what you are charging. Many advertisers may look at the new eCPM ratio to compare costs of alternate media vehicles.

When you are in a clicking environment (that is, looking at CPC), you see the cost per click that has been generated by the advertisers whose ads have been displayed on your app (always within their bidding and budget constraints, as you will see later in this chapter). You can also see the number of clicks for each ad. With those numbers, you can calculate eCPM. Take the total amount of ad revenue generated and divide it by the number of impressions (page views). For budgeting and reporting purposes, advertisers are interested in CPC, because they care about the clicks, not the impressions. However, by dividing the total revenue by the total number of impressions, it is possible to come up with a CPM that can be compared to that for other media.

■ ■ ADVERTISING CAN HAVE CUMULATIVE EFFECTS: BEFORE PEOPLE TAKE NOTICE

People receive so many messages of so many different sorts that it can take several exposures before an ad penetrates a consumer's consciousness. A great deal of market research has been devoted to specific ad campaigns and levels of awareness; marketers often have an idea of the average number of times an ad has to be seen to be effective. (In advertising jargon, this is often referred to as *frequency*, and the estimate varies by product, ad, and a host of other factors.) This can mean that an advertiser wants a specific ad to be presented to specific people more than once but not enough to become annoying.

When it comes to apps, you can use the statistics from ad networks to see not just how many users you have, but how many unique users there are. A large group of returning users may be just what certain advertisers want in order to present their message a number of times to the same people. An app that attracts many people for a single use or a rather infrequent user once a week may never be usable to an advertiser who wants repeat impressions.

■ ■ ■ How App Ads Work

Although ads on apps are a relatively new phenomenon, ads on Web pages are just about as old as the Internet. A number of players are involved with app ads:

- **Advertisers.** These are people who pay money to have their ads placed on apps.
- **Publishers.** These people create the apps on which the ads are placed. Whether you are an app developer, owner, or promoter, in the world of app ads, you are a publisher.
- **Ad networks.** These highly automated systems bring advertisers and publishers together. They collect the money from the advertisers and remit it (less a commission) to the publishers. You do not need an ad network, as you will see in the following section.

These distinctions are quite clear, but be aware that they are not mutually exclusive. One important way to promote your app is to advertise it. Thus, you can be both a publisher and an advertiser.

■ ■ INTRODUCING AD NETWORKS TO YOUR APP

Ad networks exist to bring advertisers together with apps and websites that can display their ads (publishers). Typically, these networks have two separate programs—one for advertisers and one for sites and apps that host ads. For example, Google has AdSense, which is its program for publishers, as well as AdWords, its program for advertisers. Another network, AdMob, differentiates between its two programs simply by using different areas of its site for publishers and advertisers. Now that AdMob is owned by Google, this interface may change.

You can find other networks by searching the Web as well as by monitoring and searching developer discussion groups. Facebook's developer forum has a section devoted to advertising. In this and other forums, look for the posting dates. The online advertising world has changed substantially in the last few years, and a few of the kinks that appeared at the beginning have been worked out.

■ ■ ADVERTISING ON FACEBOOK APPS

Advertising on Facebook apps is a rapidly changing area. Facebook constructs the frame of each page; when a user is running a Facebook app, the app—on its own server (yours)—supplies the content for the center of the page, and then Facebook sends the integrated page down to the user's browser. Facebook has a number of advertising programs that target ads to its customers using some of the data in Facebook's database. These are the ads that appear in the frame. If you place ads within your app, you must abide by Facebook's terms of use and terms of service. Cubics (http://publisher.cubics.com), one of the major ad networks, focuses particularly on social networking apps such as those on Facebook. Other networks may provide similar services, and there are persistent reports that Facebook itself will be entering this part of the market. Use the Facebook Developer App to get to the developers blog and discussion board, where you will find the latest information and discussions on these topics.

■ ■ WHAT MAKES AD NETWORKS RUN (AND HOW THEY PAY YOU)

You can sum up how ad networks run in one word: automation. You pick an ad network and create an account with it. If you are an advertiser, your account registration includes facts about your business as well as information as to how you will pay for your ads. If you are a publisher, your account registration includes facts about your business as well as instructions as to how revenue should be paid to you. From that point on, just about everything is automated.

Part of the automation includes a tremendous amount of training and information you will find available online, in downloadable documents, and in other forms such as webinars. Ad networks want you to advertise or host advertisements, because without you—as publisher or advertiser or both—they have no business model; that is why they provide so much excellent training for you for free.

An advertiser can generally pay based on the number of times an ad is displayed (CPM) or by the number of times people click on it (CPC). The advertiser makes that choice for a specific ad or campaign. The ad networks usually work on a bidding basis. Whenever a page is about to be delivered, the network locates all of the ads that are targeted to the type of page or app that is being delivered or run. This is based on content for Web pages and on categorization for apps. Next, the network's software evaluates the various bids and places the winning ad on the page. If more than one ad can be placed on the page or app, the process is repeated for each available slot.

In addition to the bids, there is another consideration; the publisher can usually block ads for competing brands. As noted previously, you may be both an advertiser and a publisher, so if you are hosting ads on your app, you may want to block those from competing apps.

Note that this is only a broad overview. Each network's process is slightly different, and some parts of the process (such as the way in which competing bids are ranked) are proprietary. For example, here is part of Google's criteria of its Quality Score for AdWords ads:

- The ad's past performance on this and similar sites
- The relevance of the ads and keywords in the ad group to the site
- The quality of your [the publisher's] landing page
- Other factors

This is a common approach: the scoring of ads (and bids) depends on factors related to the ad and to the page or app on which it may appear.

Notice the third point: the quality of the landing page. An ad is more likely to be shown on a page that Google rates highly for quality. Because that means you are likely to earn more money, there can be a direct link between the quality of your page and your revenue. Similarly, the relevance of the ads and keywords can affect ad placement and consequent payments. That means you really have to focus on how your app presents itself.

■ ■ How You Can Make Money

You paste some code from your ad network onto your Web page or into your app. When the page is rendered, that code is executed and the ad network is contacted. It comes up with the ad to be shown, and the final code (usually HTML) to be inserted is returned to the app. This code displays the ad and provides the link to process a click. That click also identifies your account and, in the case of CPC ads, you are credited for the click. (In the case of CPM ads, you are credited just for displaying the ad.)

The basic mechanism is simple: a portion of the revenue received from advertisers is kept by the ad network and you get the rest. The exact method for calculating this distribution is usually not revealed to the public, advertisers, or publishers.

In general, your account is credited with the money you earn from hosting ads, and the ad network settles up with you once a month.

The mechanism is the same whether you earn a few pennies or a few thousand dollars. As the owner of an app (a publisher), your goal is to increase the number of people using your app and thereby increase the possibilities for ads to be placed on your app. You can boost your chances by improving the quality of the page and making certain it is appropriately tagged so it rises higher in the rankings for ad bids.

■ ■ How Trust Is Created

As you can see, the entire process has a number of loose ends. In traditional media, publications and networks rely on rate cards— fixed prices for an ad of a certain size or length. These rate cards can be very complex, with variations for color ads, placement on an inside magazine cover, time during a network's most popular sitcom, and so forth. In addition, various discounts for volume purchases are often available. All of this changes in the online world.

As an advertiser, you create your ad and specify how much you are willing to spend for a click or an impression. The ad network's software will attempt to place your ad as many times as possible at the lowest possible cost; the amount you specify is your maximum bid. In addition to setting your maximum bid, you can also specify a budget limit for a given length of time, such as a day. As a publisher, you do not know how much people will pay for the ads that appear on your pages or apps.

Because many app owners find themselves playing both roles, you need to understand how ads work. In fact, you may decide to take a certain percentage of your ad revenue and plow it back into advertising to promote and expand your app's user base. (This has the added advantage of helping you understand the process from both sides as you evaluate your own ad's performance on other apps.)

Since the entire process is automated, ad networks provide you with a plethora of online reports, charts, and tables, whether you are an advertiser or a publisher. As an advertiser, you will not see who is viewing or clicking on your ads, and you usually will not see where those ads are placed, but you will definitely see the number of impressions and clicks.

This degree of automation means that once the ad network has invested in its infrastructure, adding another advertiser or publisher is essentially free, although there is a slightly larger load on its servers. Thus, as a small advertiser and/or a small publisher with perhaps just a blog or a single app, you can use the same basic tools as national advertisers or publishers do.

As you will see from the ad networks' sites, there is one additional consideration to this. There are links for publishers and advertisers to click on, but there also are links for agencies and national brands to use. The underlying process is the same, but large advertisers certainly get additional support.

■■■ Working Without an Ad Network

You do not need an ad network. Nothing prevents you from selling ads yourself, collecting the total revenue, and running the rotation software on your app to display the ads. For many people, using an ad network is worth the cost because all of this work is done for them, and the range of potential advertisers is large so they do not have to knock on virtual doors to find advertisers.

However, in some cases, you do have advertisers you can approach easily. These may be companies with which you already have relationships, or they may be your own company (in which case, the ads are referred to as *house ads*). The main issue you need to address is whether or not a sufficient number of ads will be available. If your app has a large number of repeat visitors, they may get tired of the same ad all the time, but if that is not an issue, managing your own advertising gives you the greatest amount of control.

It is also a good way of working with local or regional advertisers. If your app is targeted to a local audience, the restaurants, furniture stores, and car dealers in your area may not be using ad networks for apps. This depends on the region. You can usually make more money from a steady local advertiser than from distant advertisers who are relying on keywords. However, monitor the situation periodically. Local advertising is well known to be a major source of revenue if only it can be managed properly. The role of an ad network in your community might be played by a newspaper that is savvy about new technology. In that case, you might profit from using its resources to place your ads rather than, in effect, setting yourself up as a competitor—unless that is your business goal.

■■■ Don't Sell Ads: Sell Your Apps

When you publish ads on your app, you are selling that advertising space either directly or through an ad network. You have another choice: just sell the entire app. There are two main scenarios here.

■ ■ Selling the App Directly to an Advertiser

Often this scenario plays out even before the app is developed. It may be initiated by a developer or an advertiser, but the result is a custom-made app that combines a developer's app-building skills with the advertising needs of a client. If you are the advertiser, think about what your budget is and what you want to accomplish. You may be better off taking your money and paying a developer to develop a game or useful app that prominently features your brand.

Apps that are time-dependent or event-driven are particularly good candidates for this type of sale. Whether it is the opening of a Hollywood blockbuster or an appearance by a local band, a custom-written app can fit into an advertising budget no matter how large or small it is.

■ ■ Selling the App to an App Manager

Managing an app that generates significant ad revenue means tracking downloads and installs of the app and marketing the app itself so you keep your user base big enough to interest advertisers either directly or through an ad network. These skills can be learned, and many developers have done so. They then wind up with marketable skills. You can take advantage of their skills.

Ads on an app can be an important source of income, but for many, that revenue is fairly low. Nevertheless, the fact that it can be reliable, consistent, and automated means it can work for you, particularly if your app commands large audiences.

But other direct revenue streams can be truly significant. Instead of selling ads on your app, why not use it to sell products and services? Apps can make a great front end to existing e-commerce sites, and they can open up opportunities for sales channels you haven't thought of yet.

Find out more in Strategy #3.

13

Strategy #3: Sell on Your App

MANY PEOPLE VIEW THE Internet as one great shopping bazaar, so why shouldn't your Facebook or iPhone app be part of that bazaar? "Let's sell our Adirondack maple syrup on iPhone and Facebook apps" may seem like a no-brainer, but you need to give a lot of thought and creativity to the idea before you open your store. The good news is not only that can you make money from sales on your apps but that one of the biggest challenges is choosing among the many ways available to make that money.

One of the reasons there are so many ways to sell on your app is that you are, in effect, a guest of iPhone or Facebook: you are constrained in what you can do in those environments. (iPhone Web apps do not have these constraints.) The iPhone and Facebook app environments place limits on you, but the flip side is that they enforce not only a consistency in the selling environment, but also a sense of trust on the part of the user. Apple and Facebook safeguard their reputations carefully so that people will feel comfortable not just using iPhone and Facebook, but also purchasing goods and services through them. Many consumers are leery of purchasing from a website they do not know, but the major Web marketing brands are considered safer.

The following choices for you to consider can only be made after you have set your priorities and strategies, particularly exactly how you will complete sales transactions. Where the sale takes place may play a big role in your decision. If it is through the iTunes App Store, you have only one option: an iPhone app. If it is through your existing website, Facebook and iPhone Web apps are also available. iPhone Web apps are subject to almost no constraints other than the laws governing e-commerce.

In the following pages, you will see how you can take advantage of great opportunities as you start thinking about adding e-commerce to your own app—and vice versa. If you already have e-commerce in place, add an app to it. The trend today is clear: do not limit your sales to a single channel. Since the days of the first Sears catalog in 1888, retailers have realized they have to accommodate every possible marketing sales channel, including the telephone, the Web, and now apps.

■■■■ Shopping Versus Buying

In a bricks-and-mortar store, the transition from shopping or browsing to purchasing is minimized (and that is no accident!). At the critical moment, you have to reach into your pocket or purse to pull out your money, but that is a relatively easy task. In recent years, key fobs and credit cards that can be waved at or swiped through receivers on cash registers, gas pumps, and other devices have made the process even simpler, and you can rest assured that many people are thinking of ways to simplify the transaction even more.

On the Web, shopping is often easier than it is in the real world: users can compare prices easily and browse many different products and stores in much less time than it would take in person. Apps can make the experience even easier. Instead of a user deciding where to click and search, an app that is targeted specifically to the area in which the user is interested can automatically present relevant information and suggested products. Of course, this can also happen with websites, but the personalization of app interfaces on Facebook and iPhones can make apps a more effective medium.

In the end, it all comes down to the sale itself. And there you have several choices with apps.

■ ■ IN-APP PURCHASES

With the release of iPhone OS 3.0, Apple has given apps the ability to sell products that users can purchase through their iTunes account: such purchases are made though iPhone's In App Purchase. This makes it feasible for an app to pass on the handling of the payment and fulfillment functions to Apple's infrastructure. It is a valuable feature for apps, particularly those that are not part of an enterprise that already has an e-commerce sales structure in place.

There are significant restrictions on what can be sold through In App Purchase. Items are limited to those that can be delivered electronically through the app such as content, a functionality in the app that can be unlocked, and services such as access to certain data or a site or portal. Apple distinguishes between one-time services and services that are spread over time; the latter are called *subscriptions by Apple*.

iTunes and the App Store allow a user to access purchased items on all of the devices he or she has registered with iTunes. For purchases through In App Purchase, Apple makes nonconsumable products such as apps available on all the user's devices. If you sell a subscription that is managed by an external server, you must arrange to deliver it to all of a subscriber's devices.

Sales through In App Purchase are subject to the same commission Apple charges for sales of apps (30 percent).

On the Facebook side, people can purchase Facebook Credits ($0.10 each) with a credit card or mobile phone. These can then be used to purchase virtual gifts through apps as well as real gifts and contributions to charities. Like Apple, Facebook takes a 30 percent commission.

■ ■ HOW TO HANDLE APP TRANSACTIONS

In apps and on the Web, many users frequently experience a disconnect between the app or website and a separate payment site. Many online stores incorporate a fulfillment module that is a common piece of code—sometimes this code is customized with an individual website's graphics, sometimes it is not. If you have made purchases from a variety of stores on the Web, you have probably used PayPal and experienced the jump from the vendor's website to the PayPal site, then back to the vendor's site. People are generally not disconcerted by this, if only because PayPal is now a trusted payment environment. If the jump were from the vendor's website to a payment site for XYZ payment agency, a certain number of transactions might never be completed. Until a credit card or other account is actually charged, users can always stop the transaction by navigating to another page, closing the window, or even powering off the computer. (The number of shopping carts on e-commerce sites that are abandoned is high; studies report numbers in the 60–80 percent range for incomplete sales transactions.)

Whether or not you need to be concerned about this jump has a great deal to do with your customers and their e-commerce experiences. If you are dealing with people who make many (or even most) of their purchases online, they have very different expectations and sensibilities than people who are experimenting with their first Web or app purchase (congratulations, you're the guinea pig!). In a few years, this will not be a concern, but for now, as the world of apps and app-based purchases grows by leaps and bounds, every day brings new customers and app-purchase neophytes who may need a little extra explanation and reassurance.

■ ■ INTEGRATING OTHER PAYMENT METHODS

One way to handle the transition to the actual purchase is not to jump to a payment module or a service like PayPal but to jump to another tried-and-true method of purchase such as the telephone.

On iPhone, this is not much of a problem because the user stays on the same device.

The Home Shopping Network (HSN) iPhone app, shown in Figure 13.1, lets users browse items, keep track of them, e-mail them to friends, and watch the current HSN TV shows on their iPhones. To order, they just pick up the phone, or the next time they are online, they can order using their saved items.

The HSN Facebook app, shown in Figure 13.2, functions much the same way. You can browse items, move them to your wish list, and then click Visit Store to purchase.

In both cases, the method for processing an order is routed through the existing infrastructure of the Web and telephone. The apps do what they do best—the iPhone app integrates with live video and the device's telephone, and the Facebook app integrates with the user's friends, who can be notified of additions to the wish list.

At first you might think about redoing your app's entire infrastructure so it is as similar as possible on TV, in a Facebook app, and in an iPhone app, but that is usually a mistake. Make each platform do what it does best in its own environment (be it the Facebook social web or the iPhone contacts, telephony, and geolocation tools) and integrate them with what you already have. Reimplementing an entire infrastructure in an app is a good way to bust budgets, and it also means that when any part of the key infrastructure changes, you have to update your apps.

▪▪ PROGRAMMING A LINK TO YOUR INFRASTRUCTURE

One way of handling the integration of various environments with your app's existing infrastructure is to adapt that infrastructure so it can be accessed programmatically using an application program interface (API). The ideal time to do this is when your infrastructure is first being built, and many systems today have been built with an eye toward being able to be integrated with as-yet-unknown environments. Two of the leaders in this area are eBay and Amazon,

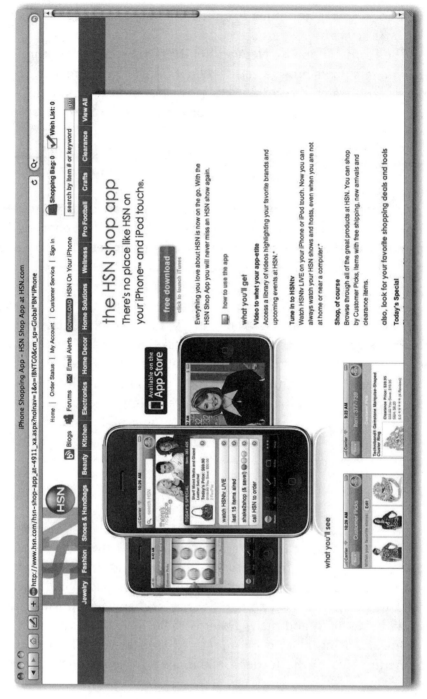

Figure 13.1

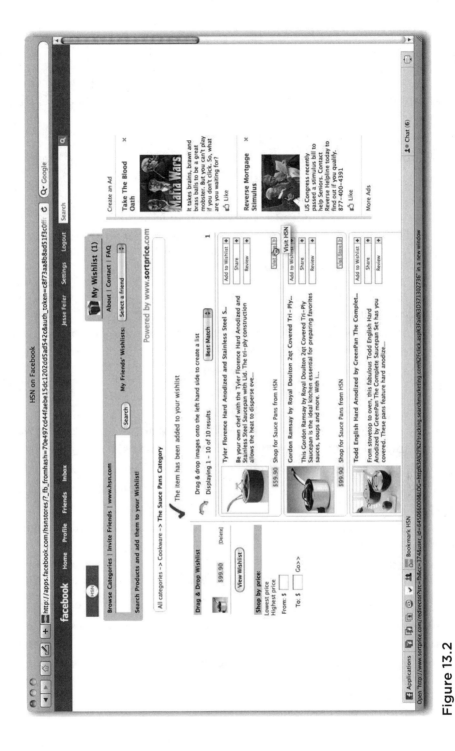

Figure 13.2

although many other companies have implemented the concept as well.

The idea is that the heart of your system can be exposed to programs that can communicate with it using code. The theory is that developers may implement their own functionality on top of yours, thereby driving more business your way.

For example, in Figure 13.3, you can see a description of the eBay Auctions Facebook app. It allows the user to search eBay, keep track of auctions, and comment on them. It integrates the Facebook social network by letting the user browse his or her friends' auctions, notify friends when the user adds or removes auctions, and share auctions with friends. Any of this can be done by writing a Facebook app that uses the eBay API.

Your job is to identify your resources and objectives in each of the app environments. Do you have a range of products for sale like eBay or HSN? If so, your objective may be to bring in customers; you should look at all the social features of iPhones and Facebook. Or are you a reseller looking for both customers and products to sell? In that case, look to the APIs of vendors like HSN, eBay, Amazon, and others so you can match customers with products.

Years ago, the notion of an API to control core processing was not widely accepted. Many programs were—and still are—designed to respond to user commands and mouse clicks. When a program is divided into interface and functionality components, it is said to be *factored*, and more and more programs are now written this way. Thus, changes to the interface can be implemented without changing the functionality and vice versa.

A good example of this is the eBay API. It was not designed to allow Facebook or iPhone apps to interact with eBay—neither existed when the API was constructed. It was developed because of an amorphous notion that people might write programs to access eBay data. In fact, in the early days, the eBay API was often used by third-party developers who combined it with Google's mapping APIs. A programmer with an idea could combine eBay and Google maps so users could search for auctions near them.

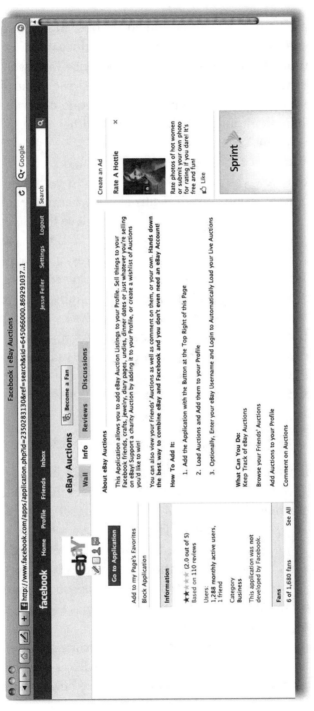

Figure 13.3

People are now realizing that the enormous, monolithic IT structures dreamed of in the past are not really achievable in a practical sense. It is much easier (and profitable) to expose whatever functionality you can to imaginative programmers and see what they come up with. Developing an app to integrate with one or more APIs can require a minimal amount of time (and a maximal amount of imagination, which is the key element).

The user does not care how you do what you do, so force yourself to think about how you would integrate an app with your possibly ancient order processing system. Think about what you would like the user to experience and what you can do to achieve that. This process is akin to asking salesclerks of a few decades ago to write up retail sales without using carbon paper. In many stores, a sale could not be completed without a yellow copy (customer), pink copy (accounting), and white copy (auditing). When it comes to apps, cast yourself in the role of the innovators who realized that the sale was important and the carbon paper was just an accessory. As you start to explore e-commerce with apps for your company, you may be accused of destroying the entire sales processing system, but you are actually liberating it from carbon paper and focusing it on sales.

■ ■ INTEGRATING WITH EXISTING CUSTOMER ACCOUNTS

Although you want to get rid of unnecessary procedures as you move to app-based commerce, make certain you retain those that are necessary. For example, in Figure 13.4, you see a description of the eBay iPhone app in the App Store. This app lets users bid on their auctions and carry out other actions, but it does not let them create an account. This is a common situation with apps. The app builds on the existing infrastructure, and that includes account creation. (The iPhone Amazon app also uses accounts that have already been created on Amazon's main website.)

There are all sorts of reasons for not allowing account creation through an app. Many people are not perturbed by an app that builds on existing accounts and doesn't let them create new ones. It fits into

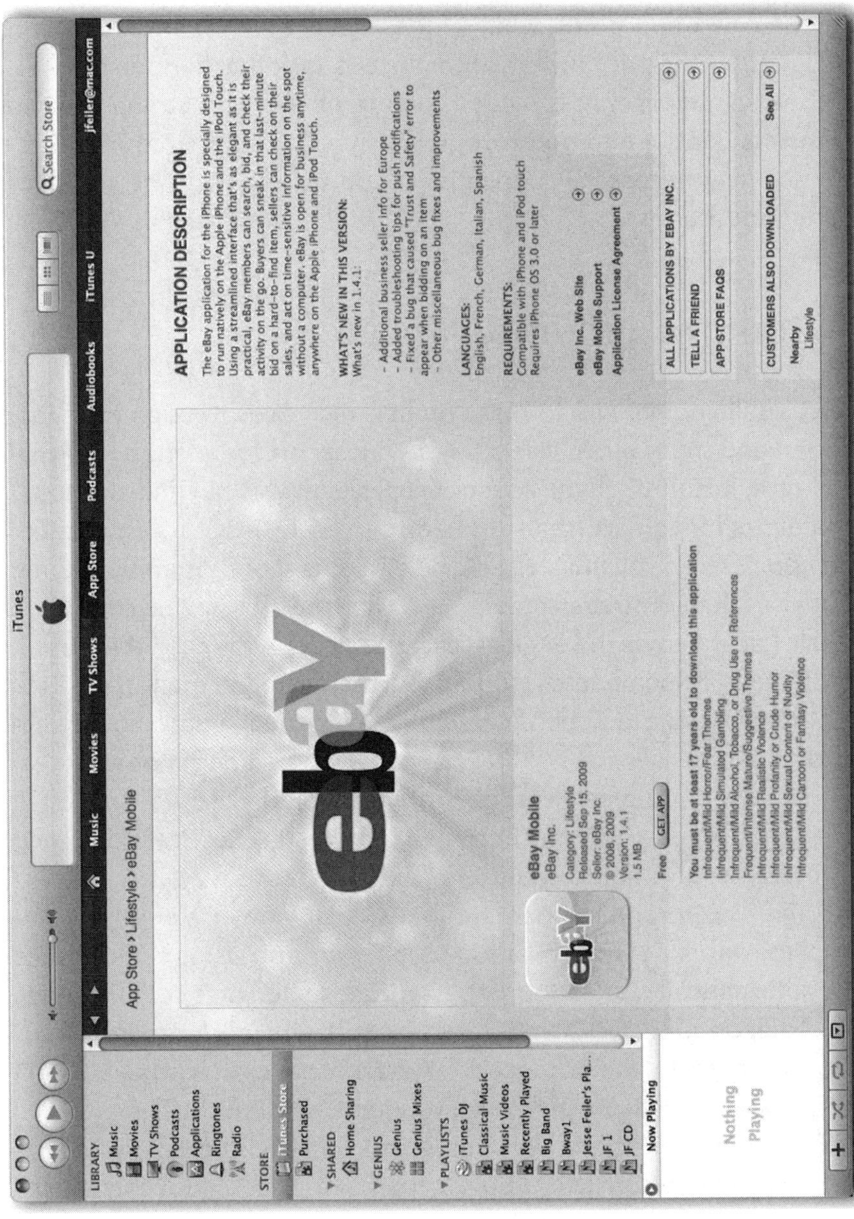
iTunes
App Store > Lifestyle > eBay Mobile
eBay Mobile
eBay Inc.
Category: Lifestyle
Released Sep 15, 2009
Seller: eBay Inc.
© 2008, 2009
Version: 1.4.1
1.5 MB
Free
GET APP
APPLICATION DESCRIPTION
The eBay application for the iPhone is specially designed to run natively on the Apple iPhone and the iPod Touch. Using a streamlined interface that's as elegant as it is practical, eBay members can search, bid, and check their activity on the go. Buyers can sneak in that last-minute bid on a hard-to-find item, sellers can check on their sales, and act on time-sensitive information on the spot without a computer. eBay is open for business anytime, anywhere on the Apple iPhone and iPod Touch.
WHAT'S NEW IN THIS VERSION:
What's new in 1.4.1:
- Additional business seller info for Europe
- Added troubleshooting tips for push notifications
- Fixed a bug that caused "Trust and Safety" error to appear when bidding on an item
- Other miscellaneous bug fixes and improvements
LANGUAGES:
English, French, German, Italian, Spanish
REQUIREMENTS:
Compatible with iPhone and iPod touch
Requires iPhone OS 3.0 or later
eBay Inc. Web Site
eBay Mobile Support
Application License Agreement
ALL APPLICATIONS BY EBAY INC.
TELL A FRIEND
APP STORE FAQS
CUSTOMERS ALSO DOWNLOADED
See All
Nearby
Lifestyle
You must be at least 17 years old to download this application
Infrequent/Mild Horror/Fear Themes
Infrequent/Mild Simulated Gambling
Infrequent/Mild Alcohol, Tobacco, or Drug Use or References
Frequent/Intense Mature/Suggestive Themes
Infrequent/Mild Realistic Violence
Infrequent/Mild Profanity or Crude Humor
Infrequent/Mild Sexual Content or Nudity
Infrequent/Mild Cartoon or Fantasy Violence
Music
Movies
TV Shows
App Store
Podcasts
Audiobooks
iTunes U
jfeiler@mac.com
Search Store
LIBRARY
Music
Movies
TV Shows
Podcasts
Applications
Ringtones
Radio
STORE
iTunes Store
Purchased
SHARED
Home Sharing
GENIUS
Genius
Genius Mixes
PLAYLISTS
iTunes DJ
Classical Music
Music Videos
Recently Played
Big Band
Bway1
Jesse Feiler's Pla...
JF 1
JF CD
Now Playing
Nothing
Playing

Figure 13.4

the idea of an app doing what it does best, which is browsing and buying rather than account maintenance (a task that is done rarely).

Creating an account is a common example of something you can have users continue to do on your website. Once the account is created and any necessary verification is done, apps can be used for future transactions.

▪▪▪ Commission and Affiliate Sales

There is still another way to make money from sales through your app. Amazon and many other sellers pay commissions for referrals. In Amazon's case, it is through the Associates program you can find at https:// affiliate-program.amazon.com. (Note the secure *https* schema rather than *http*.) The use of affiliates to handle sales is more common in some businesses than in others; for example, when the affiliate that completes the sale has expertise the original seller does not. In the case of Amazon, affiliate sales are sometimes random, but they are often made from websites that specialize in a specific topic or subject area; Amazon books are just part of a full-service site that provides information. In other cases, such as Web hosting services, it is common for affiliates to handle the actual website setup with the end users; the hosting service is designed to provide support not to end users but to the tech-savvy affiliates.

All affiliate programs operate in a similar way. For example, you can place ads for Amazon products on your app using code that is built automatically for you on the Associates site. That code includes your Amazon affiliate ID and allows people who click on links from your app to go to the appropriate Amazon page; they drag your ID along, so if they make a purchase, you get a commission.

This is similar to the way ads work, but you can do more. For example, you can construct your own graphic and text for the link, as long as you preserve the link's structure (which the user rarely sees).

Affiliate programs can provide a small additional revenue stream with almost no effort on your part. If you are already part of a program that sells products from others, think very seriously about combining your expertise and the affiliate program's products in an app that provides value in a simple package for users. If you already have the expertise and the affiliate links, much of the work has already been done—you just have to bring the pieces together and open your app for business.

■■■ Invite App Writers to Use Your Systems

What is the fastest way to move your business to the world of apps? It's not by writing an app.

Revisit your existing ordering infrastructure to see how easy it would be for an app developer to access it through an API. The worst case you may have to confront is an ancient legacy system that is structured in such a way that the user interface and the underlying functionality are intertwined. Such a system is hard to maintain, and at some point, you will have to update it. Doing so just to enable app developers to use your API may not be cost-effective on its own, but successfully exposing your API lets any developers for any devices get in on the action (and sell more of your goods).

Part of this restructuring and rethinking involves how people will access your API. Do you expose it to the world at large, or do you register developers and have them sign a nondisclosure agreement? Both strategies are commonly used.

You have to look carefully at how the API is designed—not all of the functionality should be available to third-party developers. Go over it with a fine-tooth comb with your IT staff. Do not be protective or proprietary unless it is absolutely necessary (such as for legal purposes). Can you make changes so an app writer could do the work of packaging your interface or part of it for a mass audience? Yes, the writer might make money off your system, but so would you. Many people protect their proprietary systems, but if you do not have the resources to customize every permutation of the way someone might use your system, open it up and let app writers do so. Remember that apps are about large and small numbers: a large number of sales for a small investment can make you just as rich as a single blockbuster sale.

14

Strategy #4:
Make Money with Game Apps

GAMES REPRESENT A MAJOR opportunity for making money with apps. On both iPhone and Facebook, games typically have larger user bases than all other apps. A scan of the most frequently used apps almost always consists primarily of games and, on iPhone, free apps. The same monetization models exist for games as for other apps—primarily direct sales, advertising, and indirect profits through commissions and related transactions. Games often have the additional opportunity of selling access through subscriptions and advanced versions on iPhone and bringing people from Facebook to your website for further adventures.

A Word About Gambling

Web-based games are not new. In fact, you can reasonably argue that what is now called "social networking" evolved directly from online multiplayer games. In many cases, these games cross a real legal and moral line into gambling. Although gambling is legal in many areas, and Web-based gambling games can also be legal, the opposite is also true. Many situations are clear-cut, but there are a few gray areas. This chapter addresses nongambling games. If you are interested in exploring gambling, there are many resources available, and as many people have realized, some of the same techniques that are applicable to regular games can be used to develop and promote those associated with gambling.

You have to deal with two significant issues in such cases. The first is the fact that gambling may be illegal for certain users. The second is that gambling usually violates the terms of service to which you have agreed. Those that exclude gambling range from Facebook and iPhone terms to your basic Internet connection, because many Internet service providers (ISPs) add gambling to the fine print of what you are not allowed to do.

■■■ Opportunities for Games

Explore the directories of apps on the iPhone and Facebook, and you will see how popular games are—at least among developers. Many developers get their feet wet with apps by writing games (this has been true for programs on all kinds of personal computers, not just for apps). One way to get into the app world is through a game app. Whether you write it yourself or hire someone who wants to learn, you will quickly see how the world of apps works.

■■■ Keeping Score: App Statistics Matter

Statistics matter for all apps, but they can take on added meaning for games, which often have high rates of usage (which is great for you). Your job is to stay on top of the usage so you can manage it. Do you need new features or new versions? Is your game one of the well-known fads that will be replaced shortly? If so, make sure you are the one providing the replacement. Perhaps one of the most important points to understand is that you should look at your app's stats from the perspective of your own time frame. Statistics gathered over time can provide a much different picture than daily numbers.

You can explore statistics of app use in a variety of ways. The best is to do a search for a phrase such as "Facebook apps MAU." MAU is the acronym for *monthly average users*, and it is widely used to compare apps' user bases. Because the world of apps is totally automated once the app is developed, statistics and rankings can be collected easily by automated tools. This means you can browse a lot of data to see how your own apps and others are doing.

Not all data is publicly available. For example, Apple does not normally report actual iPhone statistics.

■ ■ LOOKING AT THE BASIC STATISTICS

It is important to know how the audiences for apps are distributed on a statistical curve. The curve for games is much the same as the curve for all apps, but it may be a bit larger than some other categories.

You can find an excellent resource for Facebook app user data at appdata.com; it is part of insidefacebook.com, the Inside Facebook site. Figure 14.1 shows MAU values for the most highly used Facebook game apps (the values in this section were current as of October 2009).

The site allows you to select rankings by developer as well as by app, and you can limit the data to specific ranges of values. The listings are presented in page format. If you have the patience to go through scores of pages, you will find one that looks something like Figure 14.2. The ever-decreasing values of MAU reach 100 users at the app that is ranked 5,490, and there are still many more pages of listings after this.

The shape of the distribution is critical. Figure 14.3 shows the general shape, but it is somewhat compressed to fit on the page so the x-axis is not to scale. (The tail of the curve is actually *much* longer.) The app ranked number one in this data set has 60,322,759 MAU; the one in second place has 25,904,895. The app ranked 65 has reached a million MAU, and the one at 5,490 has reached a hundred.

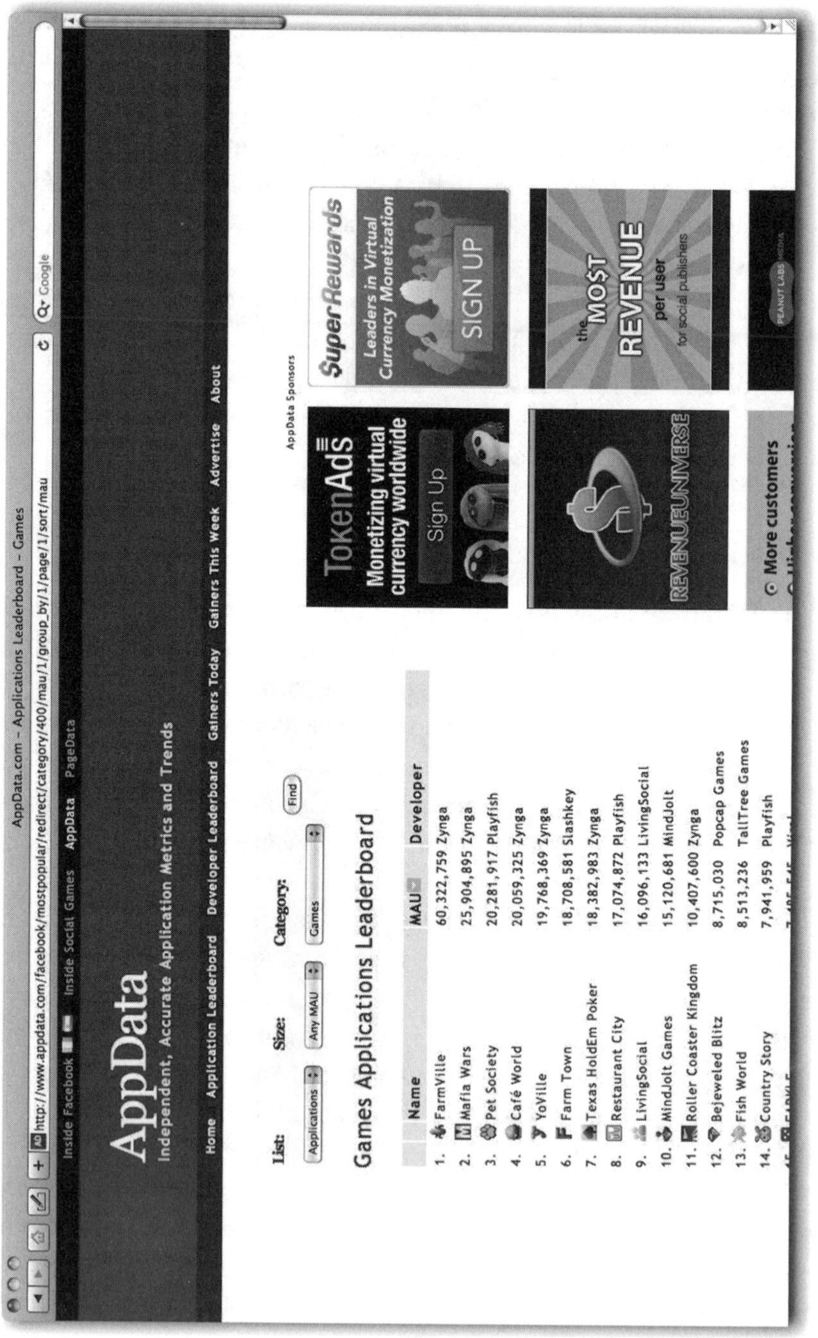
AppData.com – Applications Leaderboard – Games
http://www.appdata.com/facebook/mostpopular/redirect/category/400/mau/1/group_by/1/page/1/sort/mau
Google
Inside Facebook
Inside Social Games
AppData
PageData
AppData
Independent, Accurate Application Metrics and Trends
Home
Application Leaderboard
Developer Leaderboard
Gainers Today
Gainers This Week
Advertise
About
List:
Applications
Size:
Any MAU
Category:
Games
Find
Games Applications Leaderboard
Name
MAU
Developer
1. FarmVille 60,322,759 Zynga
2. Mafia Wars 25,904,895 Zynga
3. Pet Society 20,281,917 Playfish
4. Café World 20,059,325 Zynga
5. YoVille 19,768,369 Zynga
6. Farm Town 18,708,581 Slashkey
7. Texas HoldEm Poker 18,382,983 Zynga
8. Restaurant City 17,074,872 Playfish
9. LivingSocial 16,096,133 LivingSocial
10. MindJolt Games 15,120,681 MindJolt
11. Roller Coaster Kingdom 10,407,600 Zynga
12. Bejeweled Blitz 8,715,030 Popcap Games
13. Fish World 8,513,236 TallTree Games
14. Country Story 7,941,959 Playfish
AppData Sponsors
TokenAds
Monetizing virtual currency worldwide
Sign Up
SuperRewards
Leaders in Virtual Currency Monetization
SIGN UP
REVENUEUNIVERSE
the MO$T REVENUE per user for social publishers
More customers

Figure 14.1

Figure 14.2

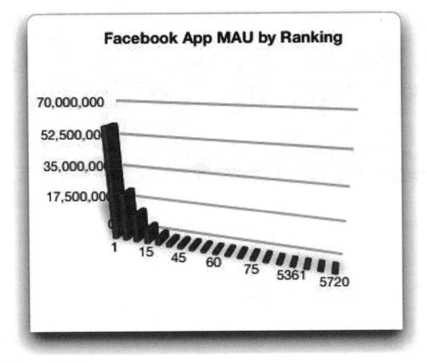

Figure 14.3

This means the highest-ranked apps have enormous numbers of users. Thousands of apps survive in the middle area. Many of them have sufficient user bases to generate a revenue stream from advertising. The stream may be modest, but because it is all automated, the cost is low once it is set up. Games are one of the categories where the less-than-blockbuster apps appear to be modestly profitable.

Similar data for iPhone apps can be found at the Mobclix site, mobclix.com/appstore/1, as shown in Figure 14.4. Both the Facebook and iPhone sites have rankings, but the raw MAU values are only on AppData.

■■ KEEPING YOUR GAME ALIVE—AND PROFITABLE

The evolution of the acronym MAU is instructive, because it started when Facebook reported daily average users (DAU). At that time, app developers tried to get the highest possible DAU scores, and many

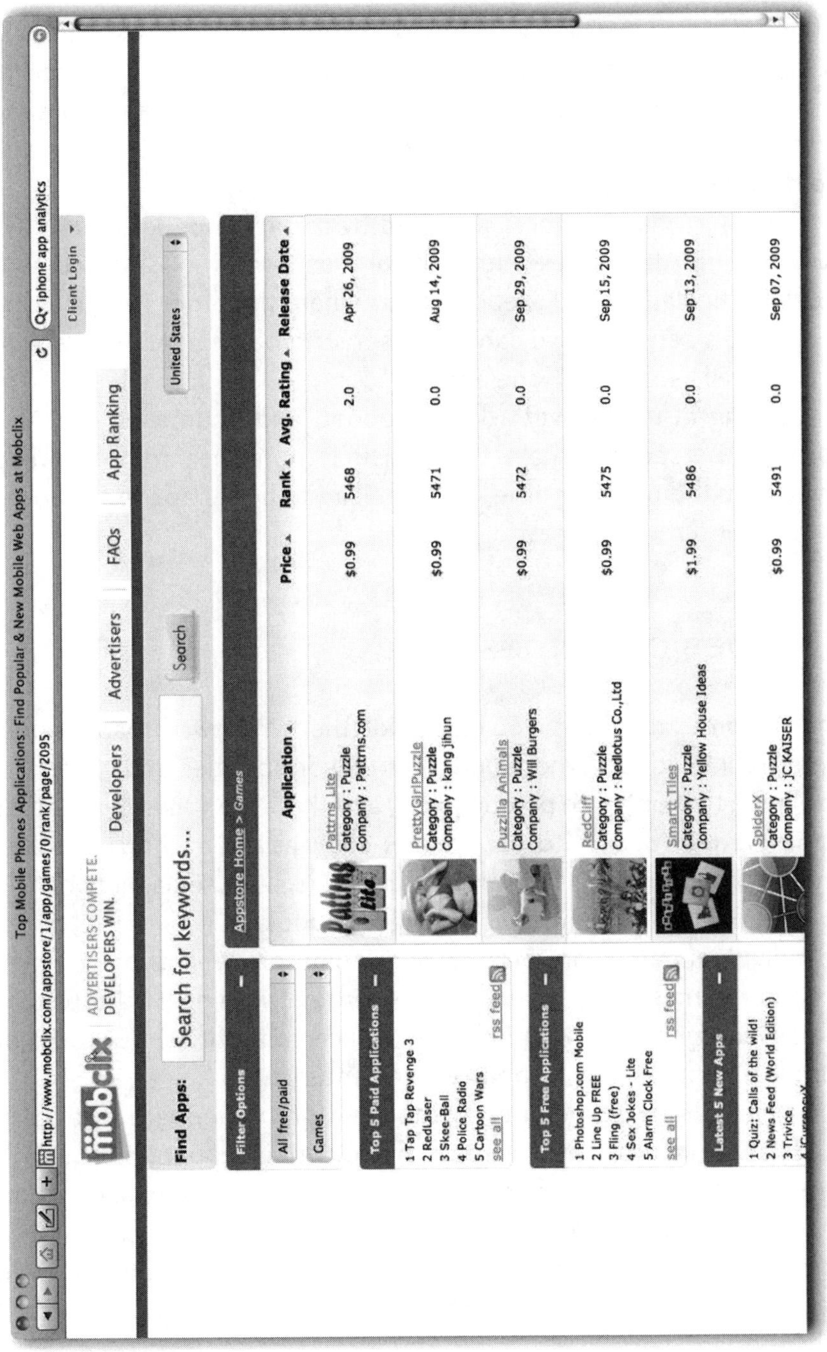
Top Mobile Phones Applications: Find Popular & New Mobile Web Apps at Mobclix
http://www.mobclix.com/appstore/1/app/games/0/rank/page/2095
iphone app analytics
mobclix
ADVERTISERS COMPETE.
DEVELOPERS WIN.
Developers
Advertisers
FAQs
App Ranking
Client Login
Find Apps:
Search for keywords...
Search
United States
Filter Options
All free/paid
Games
Top 5 Paid Applications
1 Tap Tap Revenge 3
2 RedLaser
3 Skee-Ball
4 Police Radio
5 Cartoon Wars
see all
rss feed
Top 5 Free Applications
1 Photoshop.com Mobile
2 Line Up FREE
3 Fling (free)
4 Sex Jokes - Lite
5 Alarm Clock Free
see all
rss feed
Latest 5 New Apps
1 Quiz: Calls of the wild!
2 News Feed (World Edition)
3 Trivice
Appstore Home > Games
Application
Price
Rank
Avg. Rating
Release Date
Pattrns Lite
Category : Puzzle
Company : Pattrns.com
$0.99
5468
2.0
Apr 26, 2009
PrettyGirlPuzzle
Category : Puzzle
Company : kang jihun
$0.99
5471
0.0
Aug 14, 2009
Puzzilla Animals
Category : Puzzle
Company : Will Burgers
$0.99
5472
0.0
Sep 29, 2009
RedCliff
Category : Puzzle
Company : Redlotus Co.,Ltd
$0.99
5475
0.0
Sep 15, 2009
Smartt Tiles
Category : Puzzle
Company : Yellow House Ideas
$1.99
5486
0.0
Sep 13, 2009
SpiderX
Category : Puzzle
Company : jC KAISER
$0.99
5491
0.0
Sep 07, 2009

Figure 14.4

of them did this by using a variety of techniques (many of which are no longer permitted) to drive their scores up. Facebook switched to reporting MAU on the grounds that looking at a longer period of time better reflected the user experience and encouraged both users and app developers to focus on a richer experience.

The enormous numbers achieved by certain apps in earlier days caused some difficulties with support, as server loads fluctuated wildly, particularly with Facebook apps where the processing is done on the app's server. It also can cause server problems on iPhone Web apps as well as many iPhone apps.

Looking at usage over a longer period and optimizing apps for a broader perspective make the statistics more meaningful, spread server demand out over time, and can lead to better app experiences for users and more money for you!

■ ■ ■ Types of Games

When it comes to games, you can divide them into two broad groups. There are individual games such as crossword puzzles and variations of solitaire that only one person plays, and there are social games that require several people to participate, such as online bridge. Social games fit perfectly into the world of apps, because social networking is very much the world of iPhone and Facebook.

Individual games are often variations on pre-Web and even precomputer games. They have evolved over time with new features that take advantage of the Web and new device capabilities, but most of them would be familiar to people of a century ago.

Social games sometimes break new ground. Board and card games are much as they have been since their inception, but the collaborative possibilities offered by apps and social networking can enhance them and make them much more complex. To take just one example, social games in the real world are generally limited by time and space. You can only fit a certain number of people around a card table or board game.

As you start thinking of a game, consider ways you can integrate social networking into a larger game than was possible before Web tools were available. You may start to view social networking sites such as Facebook as a new form of game in themselves. Following that path may lead to new adventures.

Notwithstanding the possibilities, remember that there is a reason why many of the legacy games are still being played. People know them and their rules, and over decades, they have provided pleasure to many users. You can put these notions together (that is, old games have withstood the test of time, and social networking allows games to function in new ways). Do some research on games that have fallen out of fashion. There is a vast horde ready to be rediscovered and reimplemented.

■■■ Categories of Games

Facebook and iPhone app directories categorize games in different ways. These categories and subcategories are likely to change over time, but those changes will become more difficult as time passes and people who are interested in game apps get used to navigating the categories to find apps they like.

Although the grouping systems have a number of overlaps, they are not identical. You may get ideas for new game apps for one platform by browsing the categories in the other. If you cannot find a category for your game app in either set of categories, consider whether you should be looking somewhere else. What you thought of as a game might belong under Lifestyle or Education. At this time, neither system of categorization allows for multiple selections, so you have to place your app where you think it will be seen by the largest number of people you want to attract.

■ ■ FACEBOOK CATEGORIES

In Facebook, games are listed in the Games category, but there is also a Just for Fun category. Each has subcategories. Browsing them may give you ideas for game apps and will show you how users are presented with app choices.

Here are the subcategories listed under Games:

- Action and Arcade
- Board
- Card
- Role Playing
- Virtual World
- Word

Just for Fun has these subcategories:

- Quizzes
- Self Expression

Both sets of subcategories have an Other listing, but the danger of such categorization is that people will not find you. Placing your app under a category or subcategory is your best chance for being noticed.

■ ■ iPHONE CATEGORIES

The iPhone Games category contains these subcategories:

- Action
- Adventure
- Arcade
- Board
- Card
- Casino
- Dice
- Educational
- Family
- Kids
- Music

- Puzzle
- Racing
- Role Playing
- Simulation
- Sports
- Strategy
- Trivia
- Word

▪▪▪ Creating Challenges

Most games let you keep score. It's true that the game's the thing and it doesn't matter if you win or lose in some contexts, but in a lot of others—ranging from the World Cup and Super Bowl to spelling bees—winning does matter. On Facebook, game scores are prime candidates for inclusion in the feed. When friends see that a user got a certain score in a game, they may be interested in clicking on the link to the game so they can try it themselves, or they may click on the user's name in the story to make a comment or catch up.

Think of this right at the start because it can influence how you keep score and what you count. You don't want to get to the point where you are ready to develop your news feed and then discover that the raw material for compelling stories has not been collected or that a critical component is missing. Which of these stories would you rather read:

- Anne got a score of 23 in the Flowers game.
- Anne is most like Citizen Kane in the What Movie Mogul Are You Most Like? game.

Arguably, this is the beginning of a kind of personality profile. Neither story is obviously better than the other for all people, but for certain types of people, one is definitely preferable to the other. Think about who your users are as you start to define stories.

On both iPhone and Facebook games, you can keep track of high scores for multiple players. This lets people compete against themselves and others. You have to be careful about publishing a high score for someone other than the current user, although it is okay to tell the user the current high score so he or she can compete with this anonymous person. Check the terms of service to see what information you can reveal through the app. You don't have to worry about the Facebook feed; users can control who sees what, so you can publish their scores if they have granted your app permission. But publishing identities and scores to people who may not be part of the user's Facebook friend network can land your app in trouble.

■ ■ ■ Expanding Brands

Apps in general let you establish a presence for your existing business in the social networking world. This is certainly true when it comes to games, but there are a few additional issues to consider. (Note that this section addresses the ways you can use apps to expand your existing game's brand. If you do not own the existing game, be very careful about infringing on the owner's copyright with your own game.) When in doubt, consult an attorney with expertise in intellectual property law.

Some games can be played equally well on apps as they can on other media. Physical games, of course, can only be simulated on an app. But card and board games can be played in similar ways; some game apps even provide ways to chat and add cross talk to the online experience.

The closer you come to the existing game, the greater your chance of cannibalizing it. If you are building an app for a card game, it does not matter that you may decrease the number of people buying playing cards (unless that is your business). With a board game, however, there are risks.

One common solution is evidenced in MONOPOLY Here & Now: The World Edition (see Figure 14.5), available from the iTunes

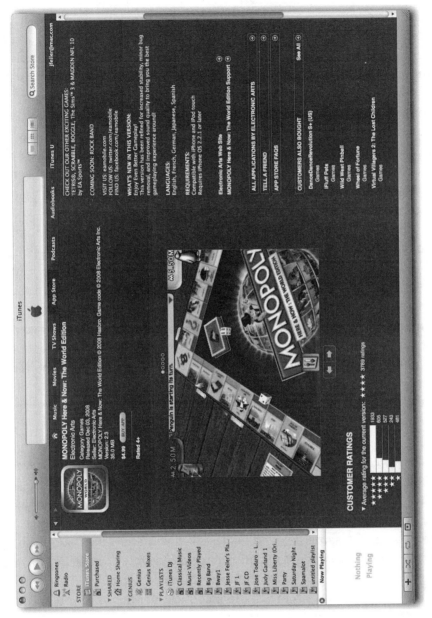

Figure 14.5

Store at appolicious.com/buy/monopoly-here-now-the-world-edition
-electronic-arts-::28621. You will discover that the app game has
slightly different features from the actual board game. It also implements a number of ideas that are impossible with the original, such
as the ability to play by yourself against the app or to play with a
group of friends on your local WiFi network. Thus, it is almost like the
board game, but not exactly. For true Monopoly devotees, this is a
serious drawback, but for most people, it is probably not even notice-
able. The ratings on the iTunes Store are very good, but if you were
to read the few negative reviews, you would see that most of them
revolve around these minor modifications.

There is another concern when you are expanding a brand to an
app, which has come up for at least one newspaper that licenses its
famous crossword puzzle. The app was originally modestly priced,
but after a while, the paper decided to move to a subscription model
so users would have to pay for a month or year at a time. (For something like a crossword puzzle that has an ongoing development cost,
the subscription model is usually best.) However, the uproar that sur-
rounded the switch to the subscription model was sizable. The lesson to be learned here is to think through your business model right
at the start. Changing prices is generally tolerated, but changing the
pricing model can be troublesome.

Games are very much their own segment of the app world. Explore
existing games to see what the patterns are, and do not fail to do
your research by reading reviews and discussion boards that evaluate
the games. (There are many of these, and eventually your game is
likely to get the same detailed analysis—if you are lucky.) Successful
games provide an ongoing relationship with devoted users, but those
users can pack up and move to the next game quickly. That's why it's
so important to stay on top of the statistics for your app: you do not
want to be surprised.

15

Strategy #5: Use Apps to Provide Your Services

APPS CAN BE A new way of interacting with your customers and potential customers. To really use them effectively, you may need to think about your operations in a different way. This chapter helps you explore these considerations and consider how you can integrate apps to provide services in a new and profitable way to customers who need them.

Provide Consulting Services with Apps

In bricks-and-mortar operations, the distinction between products and services is often not clearly defined. For example, the delivery of a refrigerator is often assumed to be part of the price, and customers in many places are welcome to save a few dollars by picking up their new appliance in their own vehicle. Some services are bundled into overhead costs, such as design consultations in home furnishing stores can be.

It is worth looking at consultation in some detail because it is such a common service that many stores provide it at a cost to themselves rather than have it paid for by revenue from customers. The model described here is applicable to a number of other consultation services ranging from financial planning to the purchase of appliances.

Once the Internet is involved, there is a different dichotomy. What is important is not products versus services, but items that can be fulfilled electronically versus those that need physical delivery. In the case of the furniture store's design consultation, it is something that can be delivered electronically. For stores that attribute design consultations to overhead and the cost of doing business, it is tempting to offer the service as part of a website. Users can experiment with furniture and accessories using interactive tools on websites or discs at no cost to the retailer once the software has been set up. (There is a minor cost to update the product images, prices, and sizes, but that is often handled with an automated data transfer from the retailer's IT infrastructure.) It can even be cheaper to provide the service on the Web because it can reduce the demands on in-store staff.

Design consultations are an obvious service to consider for apps. Some websites use old software to let users customize their choices and experiment with the look of fabrics, furniture, and paint. This software often cost quite a lot to develop, and it was frequently promoted in advertising, including in-store displays. Today, unfortunately, much of it seems clunky. All sorts of obstacles (not the least of which is often a cumbersome registration and log-in procedure) make using such design consultation tools difficult. Apps are an ideal replacement for such tools.

Think about the services you provide to customers and what your expertise is. You might want to think about questions people ask you at trade shows, via e-mail, or by phone. Is there a pattern? Can you structure the give-and-take into an app where people can navigate to their area of interest?

■ ■ ■ Opportunities for Trusted Third Parties in App-Based Consulting

In the real world, design consulting apps perpetuate a long-standing service that has traditionally been offered in stores. Alongside such in-store services, there are for-pay consulting services available from decorators and from many stores that either prefer to charge or augment their free consultations with more extensive paid services.

The chief difference between the courtesy services and the for-fee variety is the amount of work provided. In addition, there is an expectation that for-fee services (particularly if they are provided by a third party) are more neutral and objective. If you have specialized knowledge, you can market it with an app. That knowledge may come from years of experience in a certain field, but it can also come from a fresh approach. In either case, packaging it in an app can be a relatively inexpensive way to market it. (The major cost is likely to be the organization and structuring of your knowledge so that people can use it productively. This is not an app problem; it is something you will need to do whether you market your skills from an app, in a store, or at conferences and trade shows.)

Some types of consulting services, such as financial planning, are regulated so that disclosure of any payments to the consultant is required. This enforces the hope that the consultations will be objective and unbiased.

Apps can function in either model, but the benefit for consumers and retailers is that the costs of app-based consultations can be much lower than those of traditional ones. There are many opportunities for making money with app-based consultancies, but you have to be sensitive to certain issues. You can use any of the models described in this book for monetizing your apps—advertising, commissions on sales, and so forth, but you have to conform to the standards and laws that might affect your app.

For this reason, independent and nonprofit consumer consultancies such as Consumers Union and its publication, *Consumer*

Reports, have attracted wide audiences that are willing to pay for unbiased information. This, too, can happen in the world of apps and is an example of how your app might garner financial support from its users (with the general caveat that this currently applies only to iPhone apps and not to Facebook).

The product here is you and your expertise; the app is a way of making it more widely available. You may need to resist the temptation to put everything you know into your app. An app is a great way of structuring a particular part of your expertise in a particular way. Remember the earlier warning about feature creep: make your app specific and usable. You can always build a follow-on app or use the original app as a gateway to other more complex services.

■ ■ ■ How App Services Can Outperform Older Methods

Perhaps because consultation software for everything from home furnishings to financial planning has been in existence for so long, many people know what to expect from it—and many people dread their expectations. It is not just that the cost of the new living room furniture may be too high or the financial plan may show the user that he or she can retire comfortably at the age of 209, but that much of this software shows its legacies in decades-old technology. Also, some people resist asking for help of any kind, be it directions on the street or advice on landscaping. Although it is not true of everyone, many people think they should be able to do things without any help. The user control of apps can help defuse the resistance that might surface from more traditional interactions.

Using apps for providing services or any other purpose has the benefit that apps are built on modern software. If a user wants to add specific items to a wish list or make notes on furniture or financial plans, it is easy to implement. If someone wants to bounce ideas off friends, this is also a capability of apps. None of these actions needs a new design or new implementation by the developer, and users know how to use the tools.

Whether your app deals with fashion consulting, stock buying, or helping students and their parents select appropriate courses and colleges, implementing the information in an app can start from some basic platform patterns. The first is the overall design and user interface.

Service apps are most similar to games in their functionality. You need to keep a score, which is usually the cost of the project (or the projected benefit or savings), so you can evaluate one scenario against another. You have rules to play by (the size of different objects in a room, for example). For design consultations, you need the same types of interfaces found in games that let you build things (that is what you are doing, after all). For fix-it apps, the ability to rotate an object in space can be critical, whereas the ability to look at the underside of a sofa in a design consultation is likely not to matter. Or, to put it another way, gravity and "this side up" matter when you are arranging furniture in a room but not necessarily if you are rewiring an electrical outlet.

For consultations that do not involve physical objects (financial planning is a good example), you still have rules, but what exactly are you trying to move around? You can transform the consultation/ game into a physical challenge by letting the user arrange a graph or chart in exactly the same way furniture can be arranged.

Many people will say this is impossible: numbers are numbers and sofas are sofas. But when it comes to financial planning, just as many people feel their eyes glaze over thinking about numbers and arithmetic as do those who are playing with the shape of a curve.

It never hurts to take a new look at a problem, and using apps with their gamelike interfaces can be a valuable tool for you and your users.

■■■ What You Need to Know

Once you have your consultation app set up, there is essentially no cost to run it, but that does not mean that you cannot charge for the use of your expertise. Web-based consulting tools need support from a Web server. Often they use technologies that need to be retested and possibly revised each time the browsers and operating systems users run are modified.

Another advantage is that design consultation apps will be running in an environment that already supports images, such as those from the iPhone camera, and image manipulation is built into the platform software of both Facebook and iPhone. Nevertheless, there can be some implementation issues.

■ ■ CONVERTING OLD SOFTWARE TO APPS

Companies that have already invested in design consultation software can be justifiably reluctant to switch. Not only is it an additional expense, albeit it one that is usually significantly less than the initial Web-based system, but it involves retraining and developing new promotional materials. Nevertheless, it can be worthwhile, if only because both the iPhone and Facebook provide significant opportunities for promotion and viral marketing to help ease the burden of promotion for the app.

■ ■ HOW MUCH CAN YOU REVEAL IN AN APP?

Many organizations feel that consultation software requires data they do not want to export. Such information might include the actual model numbers of pieces of furniture, securities prices, mutual fund performance, and the like. If you are providing a for-fee consultation, the client pays you to collect this information and, by rights, has access to it.

If you are giving consultations as part of another service you offer, there is a legitimate concern that exposing how you do things will reveal too much to clients who may not yet have even decided to do business with you. However, in the world of the Web, more and more of this sort of information is available for free.

When it comes to any sort of planning a client might wish to do, you can add value to your app by amassing data—and so can many other vendors. You may have a significant opportunity to differentiate your services not by gathering data but by providing insight and perhaps making planning easier with your app.

This strategy just scratches the surface of the ways in which you can provide services and consultations through apps. There is a great deal of crossover between the game apps discussed in the previous strategy and the services in this one. Sometimes planning dovetails nicely with games in both apps and real life. Anything you can do to help people plan and understand the issues they are confronting can be valuable. With apps, you can create small and focused ways for people to address specific issues. Your contribution is your expertise in the subject matter and in the best ways to implement strategies.

16

Strategy #6: Click to Connect

As people explore the possibilities of apps beyond the initial "What Vegetable Are You?" (one of the earliest Facebook apps), it is important not to lose sight of what has made Facebook and iPhone apps so popular: their ease of use and their integration with the social graph, e-mail, and telephony.

It is remarkable how many developers ignore these communication possibilities and build apps that have to stand on their own without the benefits of virality. It seems somewhat bizarre that at a time when so many marketers are trying to get in on the social networking boom, some app developers are ignoring what a lot of people want. A major feature of your app for users can be the ease with which they can connect, helping the app outperform telephones, e-mail, and bricks-and-mortar enterprises. Your job is to structure your app in such a way that the interactions are simple for the user: simple to explain, simple to understand, and simple to use. Once you have a simple structure in place, add your expertise and unique perspective.

This chapter explores some of the simple built-in connections you can take advantage of when you are building your app.

■■■ Make It Easy

One of the first lessons you need to learn is that Facebook apps can go viral rapidly because it is so easy for users to take advantage of the social networking features. A single click (maybe two) can generate prepared messages, with any ad hoc additions by the user, to several friends. There is also the option of sending an e-mail message from scratch, but the major messages are simple.

This does not mean they should be canned mass-mailing messages. Facebook's platform makes it possible for developers and users to automatically send messages that are prepackaged with run-time data relevant to the current user and the actions being taken— perhaps to share information about something being browsed with a user's friends.

Nothing prevents you from implementing similar functionality on an iPhone app. In fact, you can implement many of the communication integration points from Facebook (sharing, invitations, and so forth) on iPhone. You may not have the user's network of friends to choose recipients, but you do have the user's contacts for exactly the same purpose.

On both iPhone and Facebook, remember not to generate spam, because it can get you thrown off the platform. Users should be actively involved in sending a message, and they should be able to see that message before it is sent.

You should always pay attention to the messages you generate in Facebook and iPhone apps, but there are additional complications on both platforms when you are harnessing these communication tools. If your app is a game or one of the traditional Facebook or iPhone app types, you can look at the kinds of messages that other apps in your category generate. However, if you are pushing the envelope toward a new and more sophisticated type of app, you have to deal with more complexity. It is not just a matter of notifying friends or contacts of the user's score in a game; step back and realize that you want to generate a message that comes from both you as the app developer and from the user.

■■■ Do It Right

You can use Share buttons either with default Facebook code or with code you write yourself on either iPhone or Facebook. Generally, you share a URL that is programmed behind the Share button so it is obvious to the user what is being shared (usually the article or page being viewed). Take the time to create appropriate landing pages as Share button destinations. If the Share button in your app shares information that can be found on your website, you can construct a landing page on your site that contains the information but presents it in a frame that provides context and encourages future interactions. A Click Here to Join Our List or Click Here to Get Updates button can be a valuable way to build your own viral network.

Remember that a user who has clicked the initial Share button or other link in order to share information with a friend is passing along that information with his or her name attached as a kind of reference. A message that says, "I thought you might be interested in this link" and is signed by a friend is likely to get more click-throughs than a blind e-mail (which in most cases would violate the CAN-SPAM Act of 2003). The person clicking through to the landing page is still in a known environment of friends and may be receptive to asking for more information.

Whether or not you store the contact information for the requester of shared information in your database as well as an annotation that it came from a friend reference depends on the environment and what your objectives are. With Facebook, you have to be careful not to violate the terms of service, which prevent you from storing any part of Facebook's social graph, such as the ID numbers of someone's friends. But actions that happen on a landing page on your website are not covered by these rules, so you can store any identifying information provided by the user who is clicking on your site (not on the original Facebook page). The fact that this lead was generated by Facebook (but not identifying the specific user who shared the information) ensures that you are not storing the social graph. You should always check the terms of service in the developer section of

Facebook, because issues such as this are clarified and/or expanded from time to time.

■■■ Consider the Context

Today's apps have grown out of the PDA platform—with its games, contacts, and events—and out of the full-screen computer display of the original Facebook. As apps evolve into a new form of interface that can apply to a variety of contexts, devices, and environments such as phones and game consoles, communication can go far beyond the simple "Look what I just did" and invitations that are common today.

When you set up an app for a specific purpose (be it planning a party or creating a lesson plan), users will frequently need to identify people involved by name. Not everyone is in a friends list or address book, so this is an opportunity for the user to check and see if new names should be added to these lists. By helping the user update his or her address book or friends list, the app starts to play a more important role in the user's life.

In all cases, the greater the integration of the app with a user's data (including prompting him or her to add specific data), the better the experience will be. The key here is integration. Forgo the temptation to reimplement a list of friends, an address book, or a calendar. It is amazing how many software products do this. It is sometimes the result of converting legacy software that had to implement these features because they did not exist in the environment the program was running on—perhaps twenty or thirty years ago. But often existing data stores are reimplemented either because the developer does not realize what is available or, more often, because he or she thinks the new data store requires features not available in the general one. This is rarely the case.

As always, consider your terms of service and your audience. Apps that surreptitiously store data can be the target of well-placed user ire and, in some cases, attention from regulatory authorities. Tell people what you are doing and ask permission. Do not be cagey about opt-out features: your credibility is at stake.

What is definitely OK is to let users invite friends to use your app, whether it is on Facebook or iPhone. Even if you are just asking someone to click on a friend's name and then on an Invite button, those two clicks mean the user is in control—and, of course, the friend need not act on the invitation. Do not force people to do things, and remember that users are in control—not you.

Keep in mind that when you are using built-in calendars and contacts on iPhone, you are working with a set of data that is often synchronized with personal or corporate data stores on a regular basis. On Facebook, you are frequently dealing with information that has been imported and synchronized (Facebook is not keen on exporting). But in both cases, if you can define your data set as a subset of the user's large and even shared data set that is synchronized and backed up in various ways, the experience will be better for the user.

All the pieces of app development come together as you help users take advantage of the social graph. Your app must provide value through complicated technical information, a wonderfully exciting game, or other information or services. If one user finds it interesting or useful, his or her recommendation to friends can easily broaden your app's audience. And so it goes. But remember: it starts with value.

17

Strategy #7: Let the User Forget About Saving Data

You now know about the opportunities for making money with all sorts of apps ranging from games to advertising to those that let you sell directly to customers. These strategies are not mutually exclusive, and as you think about them, consider how you can combine aspects of them into your own strategy for success in the world of apps.

Strategy #7 is about a specific type of interaction that users can have with your app, whatever it is. This is the concept of "doing the right thing"—where apps do what the user expects automatically, and that includes saving data appropriately without the user having to explicitly save documents. This feature is moving rapidly through the world of user interfaces and is flying somewhat below the radar.

Ordinary people do not examine user interfaces—how they interact with software—in great detail. They basically conclude that a program works or it doesn't—or in the case of a more complex analysis, that it is complicated or easy. This strategy builds on a feature of apps that you have probably used without thinking about it. Now it is time to think about it and about how you can profit from it.

■ ■ ■ Application Programs and Documents: The Old Model

For decades now, people have used personal computers with an interface model based on application programs such as games, word processors, or spreadsheets. Those programs let you create a specific document that might contain the moves of a game, the text of a novel, or an analysis of your household budget. When you are finished playing games, writing, or budgeting, you save your work and move on to something else. At a later date, you can run the program again and reopen your document to continue working.

Over time, many document formats became standardized, so you could open the novel you created with Microsoft Word in Pages to continue with the next chapter. Particularly in the case of specialized programs such as graphics software, you could create an image and edit it in a variety of programs. You might create a JPG or TIFF document using your digital camera, open it in Photoshop to edit it, and then open it again in Dreamweaver to place it on a Web page.

Of course, at each stage of the process, you would need to save whatever changes you made to the document so they would be there when you reopened it. *Save* is the key word here, and *reopen* is a close second.

■ ■ ■ Apps: The New Model

To use a Facebook app, you have to be logged on to Facebook. The app can query Facebook to find out who you are. iPhone apps also know who you are: each iPhone has a single user, even if several people share the device.

Apps on both environments often store their data in databases that are accessed over the Web. This makes it possible to log onto Facebook from a variety of computers and get to your account. When an iPhone app uses a database to store data, that data can be accessed from other devices that have access to the database and know the user's identification. (iPhone also has a database that is built into the device itself.) All of this enables an app to check and see who is launching it and—more important—what that person was doing the last time he or she used it. An illustration of the most intuitive feature of many apps is that if you launch an app today, it can appear exactly as it did when you stopped using it last month (or whenever you last used it). If the data is stored in a database the app can access over the Web, you can launch the app in Facebook and display the information you were working on previously, whether you used either the iPhone or Facebook incarnation of the app.

This is an app-centric world. More important, it is a world in which the old model of documents and application programs is much less relevant. You use an app, and it may be implemented on your iPhone as well as on Facebook. The app knows who you are and keeps track of what you are doing, so the next time you use it, you can just continue where you left off. The data is actually stored somewhere on the Web where it can be accessed by your various apps. This concept is called *cloud computing*.

In many cases, you do not simply continue writing the same novel or working on the same budget; you switch from one thing to another. But an app can keep track of everything you are working on using its database, and you can switch among projects as you want. And, of course, you can create new projects.

Generations of programmers have worried about implementing code to save users' data. Today, with the infrastructure that comes with apps, this is much less of a chore. The best practices for app development include automatically saving both data and the state of the app (often the last thing the user was looking at). It does not happen by magic, but making it happen is not difficult for an app developer. The end result is valuable to the user: the app just does what it needs to so the user can focus on the game or the ideas and not on how to use this strange piece of software.

None of this pushes existing technology much at all. You will find some more ideas about working in an app-centric world in the next chapter.

▪▪▪ Privacy and Security

All of this automatic saving relies on the fact that apps know who users are because users get to them through platforms that know who they are. Is this a massive loss of privacy?

The answer is a resounding "no." In fact, in many ways, people now have more online privacy than they had in the old days when people on the Internet could be identified by handles that provided no real identification beyond the chosen name.

To have robust security and privacy, the entity that provides them needs to know who users are. Facebook's privacy settings are quite detailed and allow users to set a high degree of specificity in deciding what information is available in what ways and to whom. On iPhone, users can implement the same types of security, so their information is available to them as they want it and also to people they designate.

There are two issues that can complicate this picture and dramatically reduce users' security.

▪▪ TURNING ON SECURITY

The most common way in which a user's online security can be breached is if it is simply not enabled. Sometimes this is because the user does not turn it on; other times it is because developers and vendors do not turn it on or even provide it.

In the early days of widespread Internet access in homes and small businesses, vendors typically shipped firewall software that was turned off. The reasoning was that people wanted to plug in their new devices and just get to work or play. Firewalls were an inconvenient hindrance that also happened to provide protection.

Gradually firewalls were turned on before they shipped. In part this was because people realized how vulnerable they could be with unprotected Internet access, but also because—along with everything else in the technology world—firewall management was being automated and became much easier.

Even today, most people do not adjust their Facebook privacy settings and leave them set to the defaults. As Facebook has evolved over the years, the default settings have been fine-tuned with this understanding, so the default settings provide very reasonable security, but that should not and does not prevent anyone from adjusting security to his or her particular needs.

This is not just a computer issue. You can find many stories of burglaries that were crimes of opportunity; a door is left unlocked or even wide open and—surprise, surprise—the TV is stolen.

Make certain your app earns users' trust by implementing security and selecting reasonable defaults that keep them safe even if they do not customize the settings.

■ ■ TRUSTING THE APP

An app knowing its user's identity enables it and its platforms to fine-tune security to the user's wishes. The first step is that users must turn security on and customize it if necessary. The second step is that users must trust your app and the platform to implement the security, to not reveal information they do not want revealed, and at the same time to reveal what they do want to be shown to people they choose.

If you read the Facebook terms of service or examine the rules for iPhone app development, you will see that both Facebook and Apple demand strict adherence to their privacy rules, because a major part of their credibility with users and customers comes from their protection of their platforms.

If your app violates security standards, you run a real risk of being thrown off the platform, and you may find no takers for your app. This is one of the worst offenses an app can commit. Users (and the platform on which the app runs) trust you, as the developer, with their information. You are contractually and legally responsible for guarding it appropriately.

■ ■ ■ Changing the User Interface

Making an app remember where the user was and what he or she was doing the last time the app was used raises several important design issues. One of them is demonstrated in Settings on iPhone. Settings are divided into sections such as Sounds, General, Phone, and so forth. The various settings are listed on subsequent pages. In some cases, users adjust them with a slider (as for volume), but in the majority of cases, users tap a button to move it from On to Off or vice versa.

You can search in vain for a Save button. True, when users open another app or go to Home, Settings closes automatically and saves the data behind the scenes. It is important to realize that saving data has been moved from an explicit user action to something the app just does for the user. When you think about it, having to ask an application program to open or close a document is also a task that can be automated, and apps can help with that as well.

In some contexts, certain issues such as the ability to get rid of a batch of changes (that is, close without saving) still need to be addressed, but these issues vary by the type of data involved as well as how people will use the app. Overall, the current trend is to make software do the most logical thing most of the time without user intervention. People are getting used to the world of apps. At some point, they may storm the barricades of traditional application programs and demand that they be as easy to use as apps.

Many of the so-called desktop productivity programs in use today have their roots in the 1980s, often in mainframe software from decades earlier. Some people believe that as apps provide more functionality and value with less burdensome interfaces than desktop applications, they may replace such applications in many areas.

You may want to set a goal that when that day comes (and it will), people will demand that these workhorse programs be as easy to use as *your* app. The only way this will happen is if, in planning your app, you incorporate the issues of trust and security along with intuitive interfaces so your app does the right thing for users. Remember that "doing the right thing" can be complex: you have to lead users in a certain direction so the right thing becomes obvious. Then you need to do it.

18

Strategy #8: Move Your Organization into an App-Centric World

THIS CHAPTER EXPLORES AN area that, for some people, is quite radical: moving all or part of an organization to apps. This strategy is not for every organization, and in some cases, it is not feasible. However, in other cases, not only is it feasible, it is the only way to achieve desired objectives. This chapter shows you the opportunities and various ways of implementing the strategy. And, yes, for those who are not quite ready to jump into the deep end of the pool, there are suggestions for how to implement this strategy alongside an existing website.

■■■ Defining the Problem/Opportunity

For an organization with a functioning IT infrastructure, this strategy is about moving that infrastructure in whole or in part to the world of apps. For an organization without this infrastructure (such as a start-up), it is about creating that infrastructure with apps.

The problem (and associated opportunity) is not new. The Web is terrific. Along with the Internet itself, it has dramatically changed the ways in which people live their lives. It cannot do everything, but the things it can do dramatically decrease costs and increase opportunities for people around the world. You can accomplish much of what you do using the Web with other tools (after all, people did manage to live with paper-based documents, including calendars and address books), but the Web can replace many of these older tools and create new tools for new tasks.

As the Internet has opened new doors, many organizations have found it necessary to use the Web: their customers and colleagues expect it. In some environments, older tools are still supported for a variety of reasons, so organizations have two ways of doing the same thing. (Only two, if you're lucky!)

Take customer contact as an example. Many organizations support the following modes of communication:

- Face-to-face contact at offices and retail outlets
- Telephone contact, including voice mail
- Faxes
- E-mail
- HTML forms on websites that may generate traditional e-mail or go directly into a corporate database
- Chat and text messages, particularly for customer support
- News feeds using technologies such as RSS and ATOM
- Conferences and trade shows
- Advertising and marketing through a variety of channels

The logic, of course, is that you want to make it as easy as possible for customers to communicate with your company. As you look at this list (and possibly construct a similar list for your own organization), you can see that most of the entries have evolved over time. Organizations that have been around for a long time may have started decades ago with one or two forms of communication; established firms may have been in the forefront of telephone usage a century ago.

By and large, the list grows; rarely does it shrink. In fact, so infrequently is something dropped that for many organizations the last major technology to be abandoned was the telegraph and related equipment such as ticker tape.

Although some of the media are automated (RSS and Atom news feeds, for example), the development, maintenance, and use of each one takes resources. And although there is great commonality among the various media, some of them require very different skills.

The problem in this area is that organizations are afraid that if they make communication more difficult, they will lose out on sales. The fear of being hard to reach often outweighs the obvious cost of maintaining multiple communication channels.

This does not mean that organizations do not try to steer communication into less-expensive (which usually means automated) channels. Increasingly, people notice and complain about the fact that it is harder and harder to find a mailing address or phone number for many large companies, and it is not surprising that these two labor-intensive forms of communication are being discouraged wherever possible.

The problem of multiple communication channels can be particularly serious for small businesses. Especially given the availability of the Internet and the Web, all of these channels are available to any company—even a small business run by a single, part-time owner. A totally unscientific survey of several small business owners revealed that each estimated that he or she spent at least an hour a day on e-mail and an average of an hour a day on telephone calls. (Not only was this study unscientific, but it consisted mostly of technology workers; other types of businesses would have different profiles.)

Not being able to keep up with e-mail is becoming a more and more serious problem for small businesses, because there are fewer people on whom to off-load the chore than there are at a large company. The choice for many is either to work longer hours or to bite the bullet and drop some communication channels, which is generally considered a last resort.

The first step in handling the costs associated with multiple communication channels is to recognize the problem. In some organizations, that recognition comes in one "aha" moment. This can be precipitated by a common type of event: a change to the organization name or address. Many firms would like to change their name or move, but the cost of doing so is prohibitive until it is absolutely essential (perhaps because of the imminent demolition of their office building!).

Depending on the organization, the thought of moving to an app-centric world may be something that needs to simmer in the corporate zeitgeist for a while. The cost issue is not going to go away, so it usually makes sense to start the simmering (and often to not push too hard too fast).

As noted previously, the issue is much simpler for a start-up. That may be one of the most compelling arguments for existing organizations to make the move to an app-centric world: their newest competitors will be working without the costs of maintaining multiple communication channels, giving them a competitive advantage right from the start. Some businesses can market their multichannel availability as an advantage, while others' efficient use of new media may be a selling point.

Whatever your organization's size or status, take a look at your communications infrastructure and see if you can afford to switch over to apps. The best question to ask is this: if we were just starting out (as the competition is), how would we organize communications?

■ ■ ■ Analyzing Feasibility

This is the first step, and you should take it even before you decide to move to an app-centric environment. At the very least, it will help you look at your IT infrastructure, including communications, in a new light.

Play a grand what-if game:

- What would your organization look like in an app-centric world?
- What could move where?

- What would be discontinued?
- What new features would be created?

■ ■ TRADE SHOWS AND CONFERENCES

On the national or even regional level, trade shows and conferences are often not profitable investments for most small businesses. In some places, local trade shows and conferences are created to serve the needs of the community and its organizations. The costs of participation are kept low, and the events are often arranged by the chamber of commerce or a local college.

One of the problems is that although the costs may be low, it is hard to quantify the benefits. Having more modest attendance goals than national or regional conferences reduces the effectiveness of participation for small businesses. Companies that are tied to the neighborhood—home improvement companies, for example—often do better at these shows than do companies that market to a larger area.

This is exactly where a tool such as Facebook Groups or Facebook Pages comes in, but such a tool will succeed only if a small business addresses it with the same energy it would devote to setting up a two-day display at the local chamber of commerce's annual show. You save the travel expenses and the time away from the office, but you must recognize that those savings need to be at least partially offset by your use of Facebook and your website to promote your business. If you attend several trade shows, you may be able to rethink your presence—perhaps attending some of them in person and using social tools and your website to cover for the others. Then you can integrate the two channels: your website and Facebook page will be livened up by the photos you take at the trade show. You may have to encourage friends and colleagues to comment on the photos and videos that you post (and to post their own) to maximize your company's exposure. You can keep a conversation going on the Web and on Facebook long past the short time allotted to a traditional trade show.

■ ■ WEBSITES

The websites of many organizations have grown in exactly the same way as their multiple communication channels have—gradually and without an overall plan. Many companies (particularly early adopters of the Web) feel that if they could do it over again, they would do it differently. Now is the chance to do so. Many people are moving to content management systems (CMSs) such as Drupal, which automate much of a website's management and facilitate user contributions such as comments, discussions, and uploaded photos. If it is time for a redo of your site, such a system is an excellent choice, but take a moment to also think about how you would redo your website in an app-centric world.

Every organization's needs are different, but across the board, you can find aspects of websites that are labor- and skill-intensive as well as time-consuming to develop and maintain and that may be awkward or time-consuming for your customers to use. Not surprising, these are often the interaction portions. It is difficult to retrofit the architectures of older websites to implement the sort of interactivity people expect in the world of social networking.

If you are trying to add social networking to your website, why not implement it through apps, which are the real thing. One of the hallmarks of apps is that they can interact with remote databases and websites. Some people envision a future in which interactivity is provided by apps and data stores are provided by the successors to today's websites.

■ ■ THE MAIN EVENTS: PHONE AND E-MAIL

Trade shows and conferences are a great example of how to use social networking tools to benefit a small business by providing opportunities that would otherwise not be possible. The ubiquitous communication capabilities of smartphones and e-mail are another matter: almost every person and organization today uses them. True, there are some holdouts to both technologies, but as many studies have shown, the so-called digital divide is lessening and becoming confined to specific groups. The absence of broadband Internet access in certain places is certainly a hindrance to using many new technologies, but few people doubt that this issue will be resolved within a relatively brief time frame.

If you are doing a mental exercise to see how you might function in an app-centric world, think about these mainstays of communication and how they have changed dramatically in the last few years. For example, the answering machine is unknown to more and more people; it has been replaced by voice mail that can be accessed from any telephone and often from the Internet. Websites and IT infrastructures have similarly accommodated themselves to telephones with elaborate messaging features. How much of that is now necessary for your organization?

Facebook has included built-in chat for several years now, and iPhone is obviously a major communication tool (just consider its name). Because communication is such an important part of human interaction, you have to think about e-mail and telephony carefully and somewhat differently than you do trade shows and conferences—which are somewhat marginal for most organizations—and even your website—which may be a relatively new adventure for your company and also marginal if it is not part of your core business.

In looking at the future, many people can envision a world with different forms of non-face-to-face communication (such as e-mail and telephones) than we have today. That world is becoming clearer, but it remains only a misty possibility for some people. Imagining what you want your part of the future to look like and making the transition to that concept—or to a stepping-stone to that concept—is your immediate task.

■ ■ ■ Opportunities in the New Infrastructure

The app-centric world has just as many opportunities for support services as does the current world, maybe more. Think about how many times this book has referred to your servers. An app-centric organization relies on more and different types of servers; the data centers that served large corporate users in the past still will be needed in many cases, but increasingly they may become transaction centers, reflecting the demands of apps.

Whether they are called data or transaction centers, such installations rely on the four basics: power, security, broadband connectivity, and a community that understands and welcomes new technologies. Communities around the world are recognizing that the new world of apps and cloud computing represents new economic development possibilities that can see them comfortably into the twenty-first century. (The author's hometown of Plattsburgh, New York, is one such community.)

■■■ Making the Transition

If you have thought through what an app-centric world might look like for your organization, think about how to get there from where you are now. As mentioned previously, the task is easier for a start-up because you have the option of building either an app-centric company or one that is designed to become app-centric.

Start by looking at how you might use apps in this new environment. If you are starting to build apps, these apps are the ones to build first. The first phase of a transition to an app-centric environment need not be labeled a transition: add new functionality to your organization with apps. Maybe that new functionality comes in the area of communications; maybe it comes in the use of apps for database access; or maybe it comes in the use of apps and an interactive website to expand and extend your trade show presence and cut down on travel costs. Over time, as you use these apps more frequently, you can begin to make them more central to your business, culminating in a complete transition.

Keep your options and ideas open. For many organizations, an app-centric environment will demand both Facebook and iPhones, as well as new approaches to the Web. There will surely be ripples. For example, as more and more people in your organization take advantage of the portability of iPhones, your space needs may change to reflect the fact that many people are not spending their entire workday seated at a desk. Your space needs may not diminish, but you will probably need to rethink them. This has already happened with laptops and the Internet: coffee shops, food courts in malls, and park benches replace conference rooms in some cases. Wander through a downtown area during the workday and think about how the activities you see were carried out before the days of smartphones and laptops.

Magic is not part of the transition. There may well be cost savings in an app-centric organization, but these will be offset by the cost of promoting and implementing the new environment. Facebook is a new and powerful tool for building social networks, but you cannot just rely on friends and informal games to build productive networks. You have to work at it.

Take time to think through what an app-centric world would mean for you and your organization. Even if you decide it is not right for you (yet), you should find yourself questioning many aspects of how you do business.

There is one final issue that frequently comes up. Some people will object that their customers and colleagues simply are not interested in using apps. "They're older" is one common excuse, along with "They're not technically savvy." Obviously these are real issues and apply to some organizations more than others. Still, it is hard to stop an idea whose time has come. History (particularly technological history) is littered with people who said that one revolution or another simply would not happen.

Apps are in your organization's future. They are in everyone's future in many ways. In this book, you have seen the basics of how Facebook and iPhone apps work, as well as how the app environment has taken shape over the last few (very few!) years. On the technical side, many components of apps have been around for years, but apps have brought them together in a way that has finally clicked with a wide variety of users and developers.

The strategies provided in this book are a starting point for your own plan for using apps to make money, promote your business or organization, and accomplish other goals. Use them as jumping-off points; mix and match them. The world of apps is evolving rapidly, but that does not mean you should wait for it to become yesterday's news. It is here now, and the barrier to entry is as low as it has been for almost any major technological advance in recent memory.

A long-range plan that includes apps as a core component is not a delusion—it is taken for granted by many organizations. By the same token, a long-range plan that does not include apps can put you at a competitive disadvantage.

There is really only one thing to do: get started with apps today.

Index

Note: Page numbers followed by *f* refer to figures.